**Everything You Need To Know About
Building The Custom Home**

D1122761

Everything You Need
To Know About

BUILDING THE CUSTOM HOME

How To Be Your Own
General Contractor

BY JOHN FOLDS &
ROY HOOPES

Taylor Publishing Company
Dallas, Texas

Copyright © 1990 by John Folds and Roy Hoopes
All rights reserved.

No part of this book may be reproduced in any form without written permission from the publisher.

Published by Taylor Publishing Company
          1550 West Mockingbird Lane
          Dallas, Texas 75235

Designed by David Timmons

Library of Congress Cataloging-in-Publication Data

Folds, John.
   Everything you need to know about building the custom home: how to your own general contractor / by John Folds and Roy Hoopes.
      p. cm.
   ISBN 0-87833-653-2 : $14.95
   1. Architect-designed houses—Design and construction.   2. House construction—Finance.   I. Hoopes, Roy, 1922–   II. Title.
TH4812.F655   1990
690′ .837—dc20   89–77050
                                                              CIP

Printed in the United States of America
10 9 8 7

Contents

Introduction

Although the title of this book is *Everything You Need To Know About Building The Custom Home*, a more refined title might be Residential Construction Management, because the ultimate purpose of this book is how to manage the project.

Within that objective, there are several others: to (1) help you build a *quality* home; (2) save money on building your home; (3) shorten the time it will take to build your home; (4) avoid the pitfalls and hurdles that confront any homebuilder, but especially the first-time homebuilder; and (5) prepare you for dealing with a general contractor if you opt for that choice.

We want to help you to enjoy one of the most rewarding and satisfying experiences of a lifetime—building your own home. More and more people are building homes and finding that it is not as difficult and is far more rewarding than they had imagined. Not only do you save money by being your own general contractor, but you have the satisfaction of having constructed your own environment. "It is something," Henry David Thoreau wrote in *Walden*, "to be able to paint a particular picture, or to carve a statue, and so to make a few objects beautiful; but it is far more glorious to carve and paint the very atmosphere and medium through which we look . . ."

Most people spend the greater part of their life in their home—the primary window through which they view the world. To have created it yourself, to know why a wall was built (and what is inside it) is a rare privilege. And there is nothing that will bring a family together like planning and building their own home. "It's really exciting," said one young man who had helped his father and mother build their house, "to be able to look at that wall and know all the wires and pipes that are in it, and to have stained those beams myself."

Most people think that to build a home you have to know a lot about architecture, bricklaying, carpentry, electricity, and plumbing. Not so. Most professional builders are essentially managers. They leave the actual hammering, nailing, and wiring to specialists. If you have the urge and skill to do some carpentry, painting, or even electrical wiring, you certainly can do it—(although we do not recommend it). But it is not essential that you know how to do any of the construction work that goes into a house. There are plenty of capable tradesmen out there willing and eager to help you—for a price, of course. The most important thing for you to know is not how to plumb a house, but how to get the best plumber at the right price to do it for you.

This book will help you do that. It will also enable you to control the uniqueness and quality of your house from start to finish, avoid some of the major and minor pitfalls an inexperienced builder might confront, shorten the time frame of your building project, and save money—without sacrificing quality.

The builder or general contractor on a house construction job is essentially a manager. He gets paid—usually from ten to twenty-five percent of the total cost of the house—for managing the project, not for doing any of the actual construction work. And once you grasp this concept and learn a few fundamentals of construction, there is no reason why you cannot provide the management function as well as a builder or general contractor. You don't need to dig a footing, pound a nail, or paint a board. In fact, we recommend that you do not do any of the work yourself. First, you're probably not as good as the person you could hire, and second, when you consider not only the value of your time, but what we call the "soft cost," you're going to be more expensive than the tradesman you can hire.

There are two costs in building a house—hard and soft. The hard costs consist of lumber, nails, concrete, etc., and the labor to install these materials. The soft cost is for permits, insurance, interest, and so forth. We have found that, for the nonbuilder, any hard-cost savings obtained by doing the work yourself is offset by the increased soft cost incurred.

In fact, even when you convince the lender that you're not going to do any construction work but simply serve as the manager, you still might have trouble obtaining a construction loan because of your inexperience. But there are homebuilding courses available to the average person. Sometimes the certificates you earn from these courses will convince a lender that you have the knowledge to manage the building of your own home.

We also hope this book will help you persuade a lender that you're qualified to build a house. Certainly if you completely absorb the information we provide here, you should be able to build your own house. Both the authors of this book have built a house, and one of us has spent twenty years in the homebuilding business.

1 Selecting the Land

There are more than twenty critical elements to be considered in purchasing land for building a home. Even if you already own your land, we recommend that you do not skip this chapter. It is essential to your project that you have not overlooked any building requirements. To assist you, we have provided a checklist that you should utilize prior to signing the contract to purchase your land and before going to settlement on your purchase.

It is important to note that the act of purchasing land and the act of settling on the purchase are two different things. When you purchase you are commiting yourself to the potential ownership of the land. Nevertheless, once you sign the contract, you have a certain amount of time to button down your decisions. This can be a decidedly helpful grace period, especially if you are waiting for your financing to be approved. But this period can also be tremendously stressful for some people.

Immediately after purchasing your land and prior to settlement, you will very likely experience a "rejection process" that can wreak both emotional and physical havoc. Your body, mind, and soul will try to reject your purchase. This very upsetting and real phenomenon will occur anywhere from one to six hours after you purchase the land, and may continue until such time as you actually settle on the property. It can be compared to the process of rejection that your body goes through after an organ transplant. After the purchase, you are unconsciously trying to convince yourself that you have done the wrong thing. Even your friends and family may work at convincing you that your decision to buy property was incorrect and should be reversed. But you should not be dissuaded based on either emotional factors or well-intended advice. It's natural to feel a sense of uncertainty after you have purchased anything, especially something expensive. But if there are no factual or technical reasons for not buying the property, trust your initial instincts and accept that the decision was the right one.

One way to be sure of your purchase is to know what to look for when buying property. If you've already purchased your land read through this chapter carefully; you'll rest easier if these factors have been considered. It is also a good way for landowners to troubleshoot potential problems.

We've found that people who are involved in the building of their own home end up adding more value in the final product than those people who hire general contractors to do it. The home is worth more because it is better made. It is better made because the builder is the ultimate owner; if it's your home, you'll put more tender loving care into it. Owners who are involved in their homes have a tendency to pick and choose better materials, turn away marginal lumber, and redo a wall, a cabinet, or a window if it is not just right.

CHECKLIST FOR PURCHASING PROPERTY

1. Value
2. Zoning
3. Restrictions on Land Use
4. Topography
5. Soil Condition
6. Trees
7. Moratoriums
8. Water and Sewer
9. Ingress and Egress
10. Property Access
11. Easements
12. Neighborhood
13. Environment
14. Impact Fees
15. Taxes
16. Schools
17. Insurability
18. Transportation
19. Site Preparation
20. Condemnation Proceedings
21. Variances
22. Value to You

A. VALUE

The value of a piece of property varies according to who is looking at it. You will put a value on it, the market puts a value on it, the appraiser (who also considers the market value) puts a value on it, and your lender will have his ideas about the value. Rarely will all these values be the same. It is imperative that you understand this, and be prepared for these disparities of views. But the most important measure is the value to you.

Assume you paid $50,000 for a piece of property and an appraiser puts a value of $55,000 on it. Does this bother you? Certainly not. You knew it had to be worth more than $50,000. But assume he tells the lender that it is worth $45,000. The value of variance is still only $5,000, but suddenly you decide that either the appraiser was an idiot, or you were a fool for having purchased the property. Do not despair. The appraiser has only been paid somewhere between $100 to $200 for his appraisal. You've spent weeks, months, perhaps years making a decision about the property. So in the final analysis your determination of value is not only the correct one, it's the most important one to you. Keep in mind that the whole appraisal process, despite the fact that it is referred to as an objective process, is really quite subjective. Don't be too excited if the appraisal comes in too high or upset if it comes in too low. The important thing in building a custom home, which is a very creative operation, is the value of the land and project to you. (Note: If for any reason your appraisal comes in low, later in the book we'll tell you how to deal with the situation.)

B. ZONING

Your land must be properly zoned and there are many different kinds of zoning. So it is crucial that you know the zoning regulations before you buy it. The five zoning classifications we most often hear about are agricultural, residential, commercial, industrial, and environmental.

1. Agricultural

Most people know you cannot (or would prefer not to) build a custom home on commercial property. Many, however, assume that they can build their dream home on property classified as agricultural. But this may not be true. The purpose of agricultural zoning is to provide property for agricultural purposes, and it is implicitly understood that property zoned for agricultural purposes will not be used for residential or other high-density purposes. A lot of agricultural zoning may be designed to prevent overgrowth rather than to provide for agriculture. The same is true of a relatively new type of zoning classification called environmental.

Of course, even if you buy a large tract of land zoned agricultural, you would still be able to build one structure on this property if this structure is to be your primary residence and you intend to live there and use that property for agricultural purposes. What you cannot do is buy a large tract of land zoned agricultural, start dividing that land into one-acre parcels, and then sell them off to people who want to build residential homes.

2. Commercial and Industrial

These classifications are obvious. In some cases, residential areas are changed to commercial after a home has been built on the property, although a grandfather clause would probably permit you to live in that home. But if you or subsequent owners tear it down, or it is destroyed by fire or flood, you may not be able to build another home on it.

3. Environmental

Environmental means that the land is never to be developed. The government has decided that this land should remain in its natural state—permanently. It is a step below national and state parks and it can even be privately owned. But it can never be developed. The restrictions are very severe.

4. Residential

The type of property you will be looking for will be zoned single family, residential. You have to be careful not to select a property that is zoned for duplexes. A piece of beach-front property zoned for duplexes is probably more costly than a piece zoned for single-family homes. But you cannot build a single family home on a residential property which is zoned for duplexes.

Residential zoning is becoming very complex and highly sophisticated. Counties not only classify particular areas residential, but they regulate the density within certain residential areas. The highest density would be one that permits you to build several townhouses.

Some of the more rural counties, because they have chosen not to encourage development, have been very restrictive in their residential classification. They have zoned particular areas to require a minimum of five or even ten acres to build. At one time, a county in Virginia said that you could not subdivide a piece of property into parcels of less than twenty-five acres. This type of zoning restricts people's access to the property, which some feel restricts their access to life, liberty, and the pursuit of happiness. As a result, this type of zoning has been attacked under the commerce clause of the constitution because it denies the average person the ability to own a piece of property on which he or she can build a home.

C. RESTRICTIONS ON LAND USE

When you take title to a piece of property, you do so by an instrument called a deed. A deed is evidence of fee-simple ownership in a piece of property, but that instrument does not give you unrestricted and unfettered use of your property. There are three restrictions to the use of your property: constitutional, zoning, and private convenants.

1. Constitutional

Both federal and state constitutions will dictate some of the use of your property and control some of the activities that take place on your property. An obvious control is that which deals with discrimination.

A frequently misunderstood one is the government's ultimate power to take or use all or part of your property through its power of eminent domain. Eminent domain will be discussed later at greater length.

Keep in mind that the constitution also affords you many more rights than it denies. An example would be the striking of a convenant or zoning law because it was unconstitutional.

2. Local (Zoning)

The zoning restrictions (which we have already outlined) are imposed by a local government authority that has jurisdiction over your particular piece of property. And in most cases, when we refer to the government we mean the county or the city, not the state. The state will have very little say on the use of your property. But the county or city will, and these restrictions are known as zoning ordinances. It is important to remember that the county can change the zoning on a piece of property after you have bought it. Fortunately, if there is a change, it is usually from a lower zoning classification to a higher, more valuable one, as for example, from residential to commercial—in which case, your property will be more valuable, although you will not to be able to build your home on it.

Recently, the Supreme Court ruled that when a jurisdiction downzones a piece of property it amounts to taking somebody's property rights. It was always understood that if the government took physical possession of your property—such as your land, your house, or your leasehold—they had to give you a hearing, and pay you fair market value. Until very recently, however, a rezoning of your property, although it restricted the use of your property and ultimately the value, was not deemed to be a taking of your property, and as such you were just out of luck. However, two recent Supreme Court rulings say that downzoning and building restrictions may constitute the taking of property. Hence, you have the same constitutional rights as if the state had taken your land directly—that is, you must be given a hearing and a fair market value price must be paid if the restrictions of downzoning have resulted in an economic loss.

The government (state or federal) that obtains the property through condemnation must pay you the fair market value (FMV) as determined by the court. When the government wants your property it will first try to buy it. If you elect not to sell at any price, then the property can be taken by condemnation.

When it is taken by condemnation, the court takes testimony as to value and then conveys the property to the government and requires the government to pay you the fair market value as determined by the court. The FMW, as determined by the court, may be more or less than that which was originally offered to you by the government.

The new ruling says, for example, that if you bought one hundred acres of property that you thought you could build one-acre lots on, and the county came along and rezoned it for five-acre lots, that constitutes downzoning, which constitutes a taking, and you must be given a fair hearing and paid a fair market value. This was a landmark decision by the court.

The county can also impose restrictions on your land for reasons other than building and use classifications. For example, it can prevent you from keeping animals on it or, as is often the case, require that you have at least two acres if you want to keep horses, more if you want to keep chickens, etc.

The county frequently has ordinances (called setbacks) governing the placement of your home on the property. For example, the county may require you to set back your house so many feet from the front of the property line and so many feet from the sideline. Do not minimize the setback requirements especially if you have a small, irregularly shaped, or corner lot.

You should obtain a copy of the county's zoning ordinances either prior to the purchasing of your property or prior to settlement. Be familiar with them before you commit yourself to that particular piece of property. Copies of county zoning ordinances are readily available and can be purchased from the county administrator for a nominal charge.

3. Covenants (Private)

Covenants are private rules and regulations that govern your use of your property. These private restrictions on your property are most frequently misunderstood and should be thoroughly investigated before you commit to your property and your project. One would assume that if you own a piece of property you could do almost anything you want on that property, provided it was legal and conformed to the local zoning ordinances and the laws of that particular community. But this is not necessarily true. There may be rules and regulations governing the use of your property. These rules and regulations are called covenants. These are very serious, and you should understand them fully before you purchase you land. One of them might prevent you from doing something you consider essential for the home you envision.

Some of you will purchase a piece of property that has no covenants.

If so, this section is of no importance to you except to give you an understanding of the concept.

On the other hand, many of you will purchase a piece of property (improved or unimpoved) that will have covenants. They will either be incorporated into the deed or referenced by the deed by a deed book and page number in the land records. These covenants are immutably and inextricably linked to the deed. They are the private rules and regulations that will govern your use of your property. Since they are easily enforceable and very difficult to change, it is imperative that you obtain a copy of them to determine if you are socially and aesthetically compatible with them. Because they will control how you use your property.

Covenants are imposed upon the property to control its use and ultimately to enhance its value as well as the neighboring property. In all probablility you will be buying the property, in part, because of these covenants and what they offer in terms of how you want you and your neighbors to live.

The important point here is to identify whether there are covenants on your property and if so, to determine if you and they are compatible.

Covenants are private restrictions imposed by previous owners of the property. The courts usually stay out of covenant enforcement, unless they are proven unconstitutional or unreasonable because of a changing society. Generally speaking, if the covenant is silent concerning how long it runs, it runs forever. Sometimes you will see in the covenants a particular rule or regulation that has a time period. It says this particular covenant will last for thirty years and then it will die. The words may be exactly like that. Sometimes, covenants put in by a developer will last only until such time as a certain number of lots are sold in a subdivision—at which time, it is assumed, a property owner's association will be formed and the developer will be out of it.

a. COVENANT SOURCES

(i) From the Seller
A covenant may be imposed on a single piece of property by the seller. For example, a major oil company owns a piece of property which is zoned commercial and as such can be used for a gas station. They decide to sell this property to a chicken franchise. They may sell it to the chicken franchise but they may add a covenant to the deed that says that this property can never be used for a gas station.

Other examples of covenants from a seller would be that no liquor could ever be sold from the property, that the property could not be built upon for five or ten years, or that the seller of the property would have to approve the home design before the owner could build. These covenants are perfectly legal and perfectly binding. These are restrictions that have been imposed upon the property by the seller and it will inure to either the benefit or detriment of all subsequent owners of that property. A covenant is inextricably, immutably, and permanently affixed to the property.

(ii) From the Developer

The second type of covenant, which is the one custom-home builders will most likely see, is usually part of a group of covenants, frequently called protective covenants that govern a group of lots or a whole subdivision. For example, a developer goes out and purchases one hundred acres of property and the property has no covenants on it. In fact, it is zoned agricultural. The developer then has the property rezoned to residential. At that point, the only rules and regulations governing the use of the property are the constitutional regulations and the restrictions imposed by the county zoning ordinances. However, the developer decides to impose even greater restrictions, greater protections, and greater rules and regulations governing the use of the property. As he develops this property (to be sold to subsequent and future property owners), he establishes what we call protective covenants. These covenants apply to all the lots in the subdivision, and will inure to the benefit or detriment of all subsequent property owners. The purpose of the covenant, of course, is to establish a certain uniformity within a subdivision.

The developer ends up with fifty two-acre lots. He then prepares and records with the county covenants that apply to all of the lots. If a person buys one of the lots and then sells it to a subsequent owner, who, in turn, sells it to a third and fourth person—everyone down the line will have to comply with the rules and regulations added by the developer.

b. TYPES OF COVENANTS

There are two types of covenants, affirmative and negative, and they can go on and on. An affirmative covenant might state that if you are going to build in Happy Acres your house must be a minimum of 2,500 square feet, you must put up a picket fence, or it must be a brick home. These are affirmative covenants. A covenant may prohibit horses, wood houses, modular houses, outside antennas, satellite dishes, or outside clothes lines. These are called negative covenants. Whether a covenant is affirmative or negative is immaterial; they are equally binding.

c. COMPLYING WITH COVENANTS

Do not under any circumstances think you can ignore covenants. They can be changed with great difficulty, of course but they are very binding and enforceable. And there are a number of individuals or groups that can and will legitimately charge you with violating a covenant.

d. ENFORCING THE COVENANTS

The first thing to note about enforcing covenants is that the county is not the enforcer. The county will enforce its own zoning and setback restrictions, but not the property covenants. These are usually enforced by two organizations—the property owner's association and the neighborhood architectural review board.

The second thing to note is that they must be taken seriously. A lot of people try to get around them with Perry Mason-type technicalities, but any attempt to circumvent covenants will generally be met with failure. For example, if the covenant says a house must have 2,000 square feet in it (excluding the basement) and somebody finishes off a walkout basement and argues that it is part of the square-foot living space, it will be (and has been) ruled a covenant violation. Generally, covenants are simple, straightforward, and comprehensible, so don't try to be overly creative in getting around them.

(i) The Property Owners' Association (POA)

In most subdivisions, a Property Owners' Association will probably be elected by the owners and the developer; in fact, you yourself may end up a member of the POA. Before the subdivision is fully built, the POA is usually dominated by the developer, because if there are fifty lots (and fifty votes on the POA) and only thirteen of them sold, at least thirty-seven votes will be controlled by the developer, who still owns most of the lots. As more lots are sold and more homes are built, the developer's control decreases. The POA has the responsibility to administer and enforce the covenants. The property owners do so collectively as an organization. The most likely confrontation concerning a covenant violation will come from the POA. But the POA is not the only one that can enforce the covenant.

(ii) Individuals

Any individual in your subdivision can challenge your violation of a covenant. He does not have to be a member of the POA nor does he have to be your immediate neighbor, but he must be a landowner in the subdivision covered by the covenants.

Here is an example: The covenant says no outside wood chimney chases. But one owner builds a house with an outside wood chimney chase and then sells the property to another individual, who buys the home not knowing that the wood chimney chase is in violation of the covenants. Later on, the POA or another individual sees the chimney and tells the homeowner that his chimney is in violation of the covenant. Can the home-owner say: "Well, I didn't know that when I bought the home and there-fore the covenants do not apply to me?" The answer is unequivocally no. The covenants apply regardless of your ignorance at the time you took ownership to the property. And this would be true of any situation involving improvements to the home such as decks, house color, and landscaping.

(iii) Trustees

Another example of a covenant enforcement could come about if a financial institution made a loan to any of the property owners within the subdivision. That institution would then have an interest in the subdivision. If it felt that people within the subdivision were doing things that were depreciating the property value and these things were in violation of the covenants, the institution, on its own and without the approval of the property owners' association or of any property owner, could take action to enforce a covenant. It would do so by notifying the trustees in its deed of trust to sue to enforce the covenant.

(a) The Property Owners' Association (POA)

The property owners' association is the governing body of your subdivision. It is created by the covenant and it exists to manage your subdivision and enforce the covenants themselves. There are three things to remember about the POA—the Good, the Bad, and the Ugly. The Good is that generally they are there to protect you. They do such worthwhile things as negotiate with a garbage retrieval service, and with the county or a paving contractor to maintain the subdivision's roads. Usually they also have the power to collect the money to pay for maintenance and services. And if the fees are not paid, a lien can be filed against your property. If you don't pay the lien, the POA can sell your property to collect the money. That's the good part. It maintains order, conformity, and uniformity in the community, which was its purpose to begin with.

The Bad is that many people find the rules and regulations adopted by the POA too restrictive. If this is the case, you should have done your research before you bought the property. Know what the rules and regulations are before buying a property, and you might avoid having to battle with those who enforce them.

The Ugly is when groups within a subdivision adopt two different positions relative to a particular subdivision policy or issue. For example, when a personality clash develops between one or more members of the property owners' association and the lot owner. These personality clashes can manifest themselves in political arguments over POA policies and/or covenant interpretation. These battles are not make-believe. They happen and they can be ugly. Because of the POA's power the lot owner generally loses. Even when he wins the animosity that has developed lingers for a long time.

(b) The Architectural Review Board (ARB)

The ARB is an outgrowth of the POA. It is a separate entity, but sometimes the people are the same. It is the ARB who will approve your home before you build. The good aspect of the ARB is the fact that it will exercise some control and common interest over what is to be built in the subdivision. The bad? Frequently, the covenants are drawn so that they give a lot of discretion to the ARB, and it may be that your home will be denied for what appears to be extremely discretionary decisions. If the ARB disapproves of your home or anything you plan to do and their actions are deemed to be discretionary or subjective, can you appeal their decision? You can, but you'll probably lose. The only way you can win is to prove (and it is not impossible to do) that the ARB actions were totally arbitrary and capricious. There is a distinction between being subjective and discretionary, and arbitrary and capricious, and you can battle the arbitrary and capricious.

But you cannot battle the subjective and discretionary. As an example; a person went into a limited subdivision in Virginia consisting of twenty or thirty lots. He drove through the subdivision and liked the style of homes he saw. One builder had built six almost identical homes. This person bought a lot in the subdivision because he liked those homes. He then applied to have one of those homes built and the ARB turned him down. He was going to build a home essentially the same as the other six, but the ARB turned him down because they decided that they already had too many homes like the one he wanted. The ARB won. That is not arbitrary and capricious. They had a reason for their decision.

But here is another example, that does reflect an arbitrary and capricious decision. A person drives into a subdivision, buys a lot, and submits a plan to the ARB for building a home on it. The basic plan is approved, but before the person can build his home another owner in the subdivision builds an identical home. At this point the ARB tells the person that because of the identical home having been built it would not approve his home. "However," said the ARB, "if you can modify your home to the approval of the person who has built the first home, then we will approve it."

That was an arbitrary and capricious ruling and would probably not stand up in court. The ARB abdicated its responsibility and attempted to let a single homeowner decide for the entire subdivision. There was nothing in the covenants that said you could not build two homes that were the same.

So, before you settle on your loan and take title to the property, if you have any concern at all that your home may not be approved by the architectural review board, you should submit your plans to the ARB and have the plans approved in writing by the board before you go to settlement on the property. If you are turned down once you own the property, your recourse is very limited. Your only choices are to change your home plan or sell the property.

(e) CHANGING COVENANTS

Covenants, to some extent, are like the rules and regulations governing a cemetery: everybody in the cemetery has certain rights, but they cannot be changed without the approval of everyone concerned, and this is difficult to obtain.

It is a little easier to change a covenant—but not much. It can be done jointly by all the property owners and trustees in a subdivision, but they must have unanimous approval. This is usually very difficult to achieve and the more property owners there are in a subdivision, the more difficult it is. A majority of homeowners in a subdivision cannot change a covenant; one person can hold out and prevent change.

Of course, if someone decides his constitutional rights are being violated and the court backs him, that can change a covenant. Supreme Court decisions can negate covenants.

Also, the court can decide that a covenant is against public policy. Here is an example that took place in one subdivision. The covenant banned outside television antennas. This subdivision was in an area that was down in a valley, and people had a difficult time getting television reception. Individual homeowners would put up television antennas, and part of the community would side with them and the other part would not. There were many tests of the covenants and the covenants prevailed. Ultimately, public policy prevailed and the court said: "Everybody in America watches television. These covenants run contrary to public policy because people are denied the option of watching television." This is a rare case in which the covenants were overturned.

Satellite dishes are also becoming a real problem for property owners' associations. There was one homeowner who kept trying to disguise a satellite dish which was prohibited by the covenant. First, he made it into an umbrella. But the POA asked him to take it down. Then he took a picnic table rebuilt the satellite dish, and painted it to look like a picnic table with an umbrella that was bent over. He still lost. So then he dug a hole in the backyard, lowered the satellite into it, picnic table and all, and, in effect, won the battle—but perhaps lost the war.

Do not confuse the unanimous approval to change the covenants from the less-than-unanimous approval required to elect POA or ARB members, to change the amount of association dues, or to hire a security guard. These latter decisions are made possible by the covenants and generally require only majority approval.

(f) WHERE TO FIND COVENANTS

There are several sources for obtaining a copy of the covenants—realtors, title companies, or lawyers, etc. But the ultimate source is the land record in the court house of the county or the city where your property is located. They will be referenced there and before you buy—or at the very least before you settle on—a piece of property, you should locate and read the covenants. Your deed will also give you the book and page of the covenants which are recorded for your property.

D. PHYSICAL ASPECTS OF YOUR LAND

1. Topography

When you're looking at a piece of property, consider the usable area. A two-acre piece of property is frequently as valuable as a five-acre piece of property, especially if the five acres has only two acres which are usable. Also, look at the drainage. Drainage can be a serious problem. Most states now require that the surveyor delineate a hundred-year floodplain. There are special exceptions, but in most cases you cannot build on a floodplain.

A topographic study of your property is invaluable, not only in terms of ingress and egress, but in terms of location of the well, location of the septic system, drainage, and the proper positioning of your home. If you have a large piece of property, a topographic survey may show you that if you just go a little bit further in one direction, you may have an even better site plan for your house.

2. Soil Conditions

There are a number of aspects of the physical condition of your land which must be considered in addition to the availability of water and whether the land will percolate, which we will consider later. If there are any questions about the physical conditions of your property, you should hire a soil scientist. For a few hundred dollars he will tell you everything you need to know about your soil.

a. Rock

Sometimes rock can provide a problem for digging your foundation and you might have to spend several thousand dollars to remove the rock.

There have even been cases where huge rocks have been left in a basement—too big or too expensive to move. If you're buying in a new area, you should ask about the rock content of the soil or perhaps hire a geologist if you suspect the worst. If there are houses in the area ask the owners; they usually know how much rock is around. For example, the stretch of land from Winchester to Harrisonburg, Virginia, looks very pastoral, like sections of New Zealand with sheep and cattle. What you see there are small outcroppings of rock interspersed among the pastureland. But if you dig down six to eighteen inches you run into very hard limestone which is very expensive to build a foundation on. On the surface, it looks like a cow pasture. So if you're suspicious, check your soil.

b. "Junky" Soil

The other side of the rock coin is the hardness of your land. How far down do you have to go to find soil that will support your footings? Do you have to drive pilings into the ground to support your house, as you often must do in Florida or if you are building near water? Usually the problem is too much sand, clay, or rubbish in the soil. In some areas you can dig down ten feet and still find a mixture of sand, clay, and muck that will not support a foundation. You have to keep digging until you find solid soil. In some areas you cannot build a basement because it will cave in.

c. Radon

Soil in certain parts of the country is known to emit radioactive radon. It is difficult to know precisely whether your property will emit harmful radon. After the house has been built, small cannisters (obtained from a laboratory) can be placed in your house for a few days and if there is radon the cannister will absorb it. When the test is done, you mail the cannister to the lab and it will report if you have a radon problem. If your land is in an area that has a radon problem, you county environmental agency will furnish information about what to do and who to contact for guidance and information about laboratories that do this kind of testing. On one occasion, we had a test done by a university lab and the report came back that we had a reading 3.6—the safe level was 4.0. The lab recommended doing nothing.

If you have radon present above the 4.0 level, there are two effective ways to deal with it. One is to generate proper ventilation to remove it and the second one is to prevent it from coming in. There are two ways radon could get into your house. If you have a basement, the radon could seep through the cracks in the basement floor. This is relatively easy to correct, as sealing the cracks usually eliminates the problem. If you do not have a basement, and your house sits on a crawlspace foundation (exposed soil

immediately under your floor system) this presents a more serious problem. If tests show a high level of radon and you have a crawlspace, you
should pour concrete in the foundation to put a seal between the soil and
the first floor of the house. The concrete is not smoothed out, as it would
be for a floor. This is called a wash. If you're building on a crawlspace in
an area known to have radon, you may want to go ahead and incur the
extra expense of a concrete wash. This will seal underneath your house
from radon hazard as well as other problems such as moisture. Homes
built on a slab (properly sealed) and on pilings or struts should not have
any radon problems.

In summary, if you expect radon in your area, plan ahead. Design
your foundation system to prevent radon from entering your home. A
second and/or supplementary solution is to add a ventilation system that
vents the radon gas from the home.

3. Trees

Most people seem to like trees, and they do add a nice setting to your
property. But they are very expensive to remove. A piece of property that
has trees on it is lovely, but you'll have to clear the trees from the building
site. When you clear the area, do not cut a tree off at ground level because
it is more difficult to remove a stump once the trunk of the tree has been
cut off. When you clear the site, include an area approximately ten feet
away from your house to permit room for heavy equipment around the
house. Trees that are within ten feet of the house are going to be a problem
anyway—either the roots are going to affect the house or the house is
going to be a problem for the trees. A frequently underestimated building
cost is site preparation and tree removal.

4. Water and Sewer—Wells and Septics

a. WATER

If you don't have access to city or county water you must determine
whether the property you are considering has water. You cannot usually
drill a well to find water before you purchase the property, so if you buy a
piece of property before a well is installed you're going to run a risk. Check
the neighborhood to see if there have been any serious problems finding
water and how deep you have to dig to find it. Don't be too concerned
about the geological considerations as they might look to you. Water can
be found on the top of a mountain just as well as near a river. You may
have to go a little bit deeper at the top of the mountain, but you will certainly not have to drill to sea level to find water. Most counties now require
that you have a well in before a permit is issued and most lenders require

that a well be dug before any money is advanced on your loan beyond the foundation.

If your property has a city or county water there are essentially three things you want to know about it.

(1) Where is the water pipe located? The closer it is to your property, the cheaper it will be to tap in to the water source.

(2) Is there a tap fee and if so, how much?

(3) Has that tap fee already been paid by the previous seller?

A tap fee is a fee that you pay to the county or to a utility for the right to tap into their water. It is not the cost of putting the pipe from your house to the water line, nor is it a montly fee. It is a one-time fee that permits you to tap into the line.

b. SEWER

The same situation applies to public sewers. If public sewer hookups are not available for your property you'll need a septic tank and field to handle your sewage. And it must be determined to the county's satisfaction that your soil will accept the sewage and be able to treat it through a natural process before the county will issue a building permit.

To test the soil to determine if a septic system can be installed, you need a perk test. It is a test to see if your soil percolates, which enables the effluents discharged from your home to be purified through the natural process. If your soil does not percolate (and some does not) then the sewage effluents that would be discharged on your property would present a health problem. Accordingly, if your soil does not perk and there is no public sewer, then you cannot build. Somewhere between no percolation and perfect percolation, you have intermediate situations of marginal soil. One very important thing to remember is that the perk test is determined by the number of bedrooms and not the number of bathrooms. A particular piece of property might perk satisfactorily for a two-bedroom home, but not for a five-bedroom home. Accordingly, if you want to build a five-bedroom home, you'll want to check the perk test and make sure that you can build that many.

Some soils will accept and absorb sewage and some will not. For the sewage to be able to perk through the soil like coffee, the soil must be porous. In Bermuda, which is on a coral reef, there is almost no percolation. Bermuda does not have a sewage treatment plant for the whole island. To dispose of the sewage in Bermuda, an individual must dig a big cavern out of the coral, put a few sticks of dynamite in the hole, cover it, and blow it up. That fractures the coral, which enables the effluent to work its way down through the coral and ultimately purify itself. But you cannot

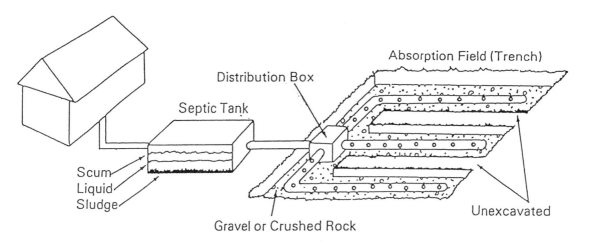

SEPTIC TANK & SOIL ABSORPTION FIELD (TRENCH)

Sewage bacteria break up some solids in tank. Heavy solids sink to bottom as sludge. Grease & light particles float to top as scum. Liquid flows from tank through closed pipe and distribution box to perforated pipes in trenches; flows through surrounding crushed rocks or gravel and soil to ground water (underground water). Bacteria & oxygen in soil help purify liquid. Tank sludge & scum are pumped out periodically. Most common onsite system. Level ground or moderate slope.

do this with all soil. You cannot fracture clay because clay will just merge back together. Sand on the other hand percolates well.

If you buy a piece of property and are told that tests have shown that it perks, you must know how long ago the test was made and what the county regulations are. If you have a satisfactory test but then decide not to build for a few years you could have a problem. The soil will not change, the county regulations may. As an area becomes more densely populated and more people move in, a county will often increase the degree of percolation required. More than one person has bought a retirement property, had it tested for percolation and found it okay, but then discovered he could not build on it a few years later when he wanted to retire. The land was the same but too many people had built there and the county percolation standards had changed. A perk test report can also lapse after time—usually after six to twelve months, depending on changes in the county health department regulations.

Since you cannot build a home without a satisfactory perk test (unless you have access to a public or private sewer system) the perk test is critical prerequisite to the acquisition of your property. Do not purchase property not serviced by a sewage system without a satisfactory perk test contingency and do not settle on your property without a satisfactory perk test. Do not accept an outdated perk test.

E. MORATORIUMS

From time to time a county will put a moratorium on building, which means it will not issue building permits until certain conditions have been changed. Moratoriums might be caused by lack of water, lack of a suitable sewer system, or inadequate facilities—roads, schools, police, or firemen. Moratoriums inspired by inadequate facilities are usually the result of rapid growth in an area. The county needs to slow down building for a while until their facilities catch up. By definition moratoriums are not permanent. But they can delay the building of your home, often for several years.

There are several types of moratoriums. You may have a moratorium on doing particular activity but still be able to build. Ordinarily a moratorium is not placed on building unless it is a broad-based moratorium. A county might put a moratorium on tapping into the sewage system. If you have a septic system then you're not prohibited from building. Or you may already have purchased a sewer tap. Sometimes sewer taps are sold in advance. The county knows the particular system will accommodate ten thousand homes and they have sold nine thousand sewer taps. The next one thousand people can tap in and you can buy the tap before you build. Someone might buy a tap for a piece of property, not build a home, then sell it to you—in which case you can still build. But if there is a sewer moratorium and your land does not perk you could face a long wait for a new sewage treatment facility to be built.

Checking for moratoriums is simple. Just ask the county administrator if there are any moratoriums or if any are anticipated, and if so, would they apply to the property which you are considering purchasing.

F. PROPERTY ACCESS

How easy is it to get on and off your land from a highway? In assessing your property, be sure to take into account the access to it. Will your access comply with county easement regulations? If you have a ten-foot drop or perhaps a twenty-foot climb before you reach that portion of your land where you want to build a house, it might cost you a substantial amount of money to build your access road. But don't walk away from a piece of property merely because the building site is difficult to get to. The price on the land may lower because of this, and once you get to your building site it could be gorgeous.

The best thing might be to have a topographical map of your property made. This could cost from $300 to $1000. It will show you how the property would feel if you ran your hand over it. It shows you the ups and downs. And it will enable you, the land planner, the architect, the engineer, and the builder to properly situate your house and properly obtain access, ingress, egress, and proper drainage. Frequently people will ignore a piece of property where less than three or four percent of the property is on the road, and what does abut the road is the least accessible and least attractive portion of the property. For a few thousand dollars, you may be able to break through that area, open it up, and thus have one of the most beautiful pieces of property in the neighborhood. Unfortunately, a lot of people make decisions based upon first sight, and that can be a mistake.

1. Ingress and Egress

If you buy a piece of property adjacent to a highway you will need to obtain a permit to create an access to the highway. Rarely will it be denied, but you must get permission from the state highway department. They will specify exactly where you can come in and where you can go out. If you're in a heavy traffic area, you could be required to have a site easement or an acceleration-decleration lane. Normally you see this on commercial property, such as fast food purveyors. If you are building in a rural area an ingress-egress permit is especially important. But even in a subdivision you do not automatically have the right to put a driveway wherever you want. You must check with the county as well as the POA or ARB.

2. Easements and Rights-of-Way

Easements and rights-of-way are a necessary consideration of prospective building sites—and they are by no means an undesirable feature. An easement is a privilege, or convenience, which one neighbor has of another by prescription, grant, or necessary implication as a way over his land—a liberty or privilege without profit. An easement runs to the dominant tenant (the one who has the right to use the easement) and takes from the subservient tenant (the one over whose property the easement runs).

Let us say you have a five-acre piece of property that has been subdivided into parcel A and parcel B as shown in the illustration. In this example, the owner of parcel B granted an easement to the owner of parcel A. Parcel A is called the dominant tenant and Parcel B is called the subservient tenant. This easement will last in perpetuity, which means that later on B cannot say to A, "Time is up, you can no longer use my property to gain access to your property."

An easement is a right of use to land that crosses over B to A, and cannot be revoked. A has the right to use the access strip, but the right is restrictive—it can be used only for ingress and egress. He cannot build on it or plant on it without express permission from B.

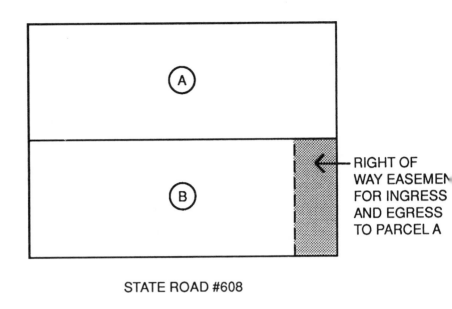

STATE ROAD #608

But there could be a problem. If the easement were obtained in 1912, and at that time only a forty-foot-wide easement was needed to obtain a building permit, today the county might require a fifty-foot-wide easement access, in which case you could not obtain a permit to build on Parcel A.

At this point you may consider purchasing an additional ten feet of easement width from the owner of parcel B. Although this may be possible, don't count on it. The owner of parcel B may not want to sell under any circumstances, or, even if he did, he might be prevented from doing so by the covenants. Fortunately, in this case you could probably get a variance from the county which would solve the permit problem. Solving the permit problem may still leave you in a bad position as the owner of parcel A. Some lenders may not make you a loan without fifty feet of road frontage. Both of these hurdles, although probably ultimately crossable, are at best, nuisances, and at worst, project busters. So it is important that before you buy or settle on your land you analyze the easement that gives you access to your property. If your property does not have state road frontage or road frontage itself you need to check the easement or right-of-way and make sure it is not only functionally usable but will permit you to secure a building permit and obtain a loan to build your home.

In addition to consideration of ingress and egress created by easements, there are other easement considerations. Not only do you want to consider the easements that benefit your property as the road access issue described above, you also should consider easements that run over your property. These easements may be for other roads, power lines, telephone, etc. You need to decide aesthetically how easements affect the property which you are considering purchasing.

3. Variances

If you find that restrictions or regulations imposed on your property by the county are creating a problem, do not feel that it is necessarily hopeless. It is possible in some cases to obtain variances from the county, although it might cost you some time and effort (and maybe even a lawyer's fee).

In most jurisdictions there are two conditions under which a variance can be granted. One is when without the variance there would be an undue economic hardship on the property owner. Such hardship must be demonstrable. Also the granting of the variance must not set a precedent. So if you have a hardship but the granting of a variance to you will set a precedent, in all likelihood the variance will not be granted. You need to satisfy both conditions in order to obtain your variance.

You cannot knowingly buy a hardship and then ask for a variance. If you had prior knowledge of the problem but bought the piece of property anyway and then asked for a variance, more than likely it would be denied.

G. SPECIAL CONSIDERATIONS

1. The Neighborhood

The desirability of a neighborhood is a subjective matter. Consider what is happening to a neighborhood, and which way it seems to be going. There is one aspect of a neighborhood that is not subjective, however, and that is the way the bank (or lender) looks at it. If you plan to build a $400,000 house in a neighborhood of predominantly $200,000 houses you will have a problem. The lender will not only deny you a coventional loan—80% of the cost of your house—it will probably not even give you twenty-five percent of the cost of your house, because if there is a foreclosure the lender will have difficulty selling it even at a price below what it is worth. This can work both up and down. So if the cost of the house you plan to build is disproportionate to the value of the other houses in the neighborhood you could have a hard time obtaining financing.

One of the biggest problems for people who build custom homes is the tendency to not only overbuild, but to overbuild for the neighborhood. A home which is significantly overbuilt for the neighborhood will not only look out of place, it will be very difficult to finance.

Of equal importance in assessing the neighborhood is the identity of your immediate neighbors. You may find a neighborhood that is acceptable to you, but you may have a neighbor who has been a longstanding problem. This can be a factor in both an urban setting and a rural setting. Sometimes you will find a neighbor who is irascible, or who exhibits anti-social behavior (loud noise, late parties, or twenty cats), but the fact may not be immediately apparent when you're buying a piece of property. It is something that should be carefully researched, because one bad neighbor can spoil your whole project.

2. Environment

There are a number of things that could affect the environment of your property, and most of them are obvious. One that might not be so obvious is the wind. On some days, for example, the wind could blow toxins from a nearby lime factory across your land. Or the wind might be responsible for your property being plagued with an unusual number of flies, mosquitoes, insects, odors, etc. Keep that in mind as something you need to investigate. The realtor might not want to tell you, so ask the neighbors—or the children. A neighborhood child is frequently an excellent source of information about the real situation in the neighborhood. Talk to neighborhood children about what is going on there.

3. Impact Fees

Impact fees already exist in Florida, California, and Maryland, and as the population grows will inevitably spread to other fast-growing parts of the country. As this book goes to press other states are considering impact fees. Every time someone applies for a building permit that implies an impact on the community. An impact fee is a one-time charge for the economic costs associated with the impact on that community. The money is used for general revenue to fund the police, fire departments, the schools, and other services.

You pay the impact fee when you obtain a building permit, not when you buy the land. There are also some real constitutional issues now being raised about impact fees. If such fees are not used for the purpose for which they are specifically charged they probably will be ruled unconstitutional. Nevertheless, in very dense and rapid growth areas, impact fees will be a part of the building process for the foreseeable future.

4. Taxes

Taxes vary from county to county and state to state. In some areas the tax rates are so high that you might want to consider buying a piece of land somewhere else—especially if you can find something just as nice and just as convenient nearby but across a state or county line.

5. Schools

Everyone is aware of the differences that can exist in various school systems. As as a rule of thumb you will find taxes a little higher in counties that have the best school systems, but usually it is worth it—not only for your children's education, but in the event of future resale. Most people prefer to buy in an area with good schools even if taxes are higher.

6. Insurability

You will not make a decision about purchasing a piece of property based on fire insurance, but you should know that it will cost you more in a remote area. The type of home you build will also affect your premium. For example, brick houses are cheaper to insure than frame. If a frame house catches on fire in an area that is some distance from fire hydrants it is probably going to burn down. The brick house is less vulnerable to fire.

7. Transportation

Most people do not even begin looking at specific pieces of property until they are already satisfied that the transportation or commuting to that area is no problem. You should give some consideration to public transportation. There might be another oil crisis, and if gasoline is rationed or cut off for a time transportation to a remote area could present a serious problem.

There are, however, pros and cons to the presence or absence of public transportation. Some people don't want to be near public transportation because it will eventually bring more congestion to the area. On the other hand, some people prefer to live near it. With this in mind, you need to look at the county's ten-year plan for that particular area. The piece of property you are considering buying which may have a nice bucolic setting may be located right next to a road that is going to come through when the ten-year plan is completed. The projected road may even be the beltway or the perimeter highway. We have seen situations in which people bought a piece of property in a rural setting with a big span located between where they were buying and the next available property. What they failed to find out was that a major right-of-way had been acquired through the area by the county for a future highway. Information about future plans for the county is available at the county headquarters.

8. Site Preparation

What will you have to do to your property to prepare it for building? From a building point of view, the most desirable site is what builders call pancake farmland. All you have to do to prepare your site is put the gravel down for a driveway. A more aesthetic site, of course, may be the heavily wooded lot with a lot of ups and downs. But on this kind of lot it is not unusual to spend several thousand dollars preparing the site for the building. If you're building a $300,000 home it is probably worth it, but you need to be prepared for the cost. Again, a topographical survey will provide you with the information you need to determine how much it is going to cost to prepare your site for construction.

The beauty of pancake property to a builder is that you have no extraordinary site work. You don't have to haul material in or out over difficult terrain. You don't have to make special arrangements to move your equipment around the house or for the delivery of material or for access. If your property is not perfectly flat and level, however, you need to consider the cost of both access to your building site and preparing the site itself. Such costs are frequently underestimated.

9. Condemnations

Under the right of eminent domain, the government can take your land for any reasonable use. So be sure to check with the county before you sign a contract to make sure your land has not been condemned or in condemnation proceedings—not condemned because it is a bad piece of property, but because the county or state plans to build a road through it or a sewage treatment plant on it. Some people have no hesitation about selling land that has been condemned. So be sure to check with the county—especially if the land is being sold for what appears to be a below-market price.

10. Value to You

Finally, in selecting a piece of property the most important thing is its value to you. If you're planning a custom-built home, obviously there is a certain creativity and uniqueness in you, and that uniqueness should be extended to your property. Its value should be what it means to you rather than what it means to the appraiser. In the last analysis your home is your castle, and the land you build it on is the setting of your castle. The nature of your land and the type of home you build on it are the result of choices only you can make. We hope this information serves as a convenient checklist for helping you make these choices.

2 Contracts

In the process of purchasing property and building a home you will be signing many contracts and purchase orders (a form of contract). Accordingly, it is important for you to understand some of the legal aspects of a contract and the contracting process.

A. DEFINITIONS

1. Contract

For our purposes, we will use the definition of a contract taken from *Real Estate: Principles and Practices* (South-Western, 1987) by Maurice A. Unger, a professor of Real Estate and Business Law at the University of Colorado.

> Generally, a contract is an exchange of promises or assents by two or more persons resulting from an obligation to do or refrain from doing a particular act, which obligation is recognized or enforced by law.
>
> A contract may be formed when a promise is made by one person in exchange for an act or the refraining from doing of an act by another. The substance of the definition of a contract is that by mutual agreement or assent the parties have created legally enforceable duties or obligations that did not exist before.
>
> If all the terms of the contract have been fulfilled, it is said to be executed. If something in the contract remains to be done, it is said to be executory.

2. Real Estate

We also need to define what real estate is as it relates to contract law. It is very important to understand the technical meaning of the term. Real estate is land, dirt, rocks, and trees. If you have a piece of property that is unimproved, it is real estate. If it had a house on it when you bought it, both the property and the house are real estate. If you buy the property without a house on it and you add a house, the house is real estate. Real estate is the natural land and any improvements that are on it.

What real estate is not is personal property that is not part of the improvements or not yet part of the improvements.

Your fire insurance policy or your homeowner's or builder's risk policy will cover the improvement in your real estate. As you build the foundation, the policy will cover the foundation; as you build the first floor, the policy will cover the first floor; and when you build the house, the policy will cover the completed house. But, before it becomes attached to the house, the material to be used is personal property.

So, if you're building your house and someone delivers twelve windows to your job site and six of them are installed and six are on the ground, and there is a fire that night, your builder's risk policy will cover the windows in the house because they are part of the real estate. But it will not cover the windows on the ground, because they are personal property. It is possible to add a provision to your insurance policy (known as a rider) that covers building material stored on the site. Accordingly, it is important that you understand the concept of real property, the distinctions between real property and personal property, and the legal and insurance ramifications associated with the distinctions between the two.

B. TYPES OF CONTRACTS

1. Oral vs. Written

Many people think that contracts must be in writing, but they don't have to be. Let's say that you drive up to a gas station on a hot summer day, and say, "Fill 'er up," and the attendant pumps gas into your car and then says, "That will be fifteen dollars." And you reply, "No deal. We didn't have a contract." Who wins?

He wins. "Fill 'er up!" is an oral contract.

You go into a restaurant. You order a meal. "I'll take the filet, I'll take the asparagus, I'll have a coke." The waitress serves all of this and gives you a bill. Are you obligated to pay? Absolutely.

So an oral contract clearly can be a contract, and it can be as binding and enforceable as a written contract.

2. Affirmation by Action

A contract need not even be oral. You drive into the same gas station, but now it is pouring down rain. The kid comes running out. Your windows are rolled up and you signal with your hands: "Fill 'er up!" And he signals: "okay." He puts the gas in. Do you have a contract? Yes. It was implied. And an implied contract is an enforceable contract.

Similarly, if you go into a restaurant and you start eating the buffet, that binds you. It is an implied contract.

Sometimes a contract that is not a contract at the time can become a contract at a later date. For example, if you say, "Let's sign the documents now," and you don't, and then you subsequently perform as if the contract had been signed, that is called *affirmation by action*. Your subsequent actions may ratify a contract.

3. Executory vs. Executed

An executory contract is one in which there is something still to be done— and it is very important in real estate. Most contracts are executory. Something still remains to be done. When the terms and conditions of the contract are completed, it is said to be executed.

If you sign a contract to buy a piece of property, it is an executory contract because you have agreed to buy that piece of property in the future and the seller has agreed to sell you that piece of property in the future. Generally there are some conditions and terms that have to happen in the interim. You are both bound by the terms and conditions, but they are things to be done in the future. When you settle on your property and you give the owner the consideration and he gives you a deed, that transaction then becomes an executed contract.

C. IMPORTANT CONSIDERATIONS

1. Statute of Frauds

The statute of frauds is an integral part of contract law. It states that some contracts must be in writing. It does not say that an oral contract is not a contract; it says only that some contracts must be in writing. A written contract is more lasting, more comprehensible, and less controvertible than something oral. So as the contract becomes important and potentially more controvertible, the law says it must be in writing.

The statute of frauds, which comes from English law, has been adopted by most countries that follow the English legal system. It covers many aspects of contract law. In this book we will discuss only three considerations in the statute of frauds which are relevant to real estate and contract law as they relate to building a home.

(1) The statute of frauds says that all contracts must be in writing if the transaction involves real estate.

(2) In most states, the Uniform Commercial Code says that any transaction involving the sale of goods and services in excess of $500 must be in writing. If you were to go into a department store and buy something for $400 the transaction would not have to be in writing.

(3) Any contract that will take longer than a year to perform must be in writing. This is only logical.

2. Parole Evidence

Lawrence P. Simpson's *Handbook of the Law of Contracts* (West Publishing, 1965), which is frequently referred to as the Bible on contracts, says:

> When a contract is expressed in writing which is intended to be the complete and final expression of the rights and duties of the parties, parole evidence of prior oral or written negotiations or agreement of the parties or their contemporaneous oral agreements which varies or contradicts the written contract is not admissible.

The term "parole evidence" has nothing to do with getting out of prison on parole. In contract law, parole evidence is essentially any oral discussion that occurred prior to or contemporaneously with a contract. Here is how it might work in a real estate transaction: Judy has said she will sell John a house. They are close to arriving at a price but John is still undecided, so Judy says, "I'll tell you what I'll do. I'll plant pansies on your property."

John says, "I don't like pansies. They have to be taken care of."

Judy says, "I'll take care of them. I'll plant the pansies and I'll mulch them if you buy this house for $100,000."

They then write the price of the house in the contract. The contract says nothing about mulching pansies, although it does say they will be planted. They go to settlement, and John says, "Nice house, but how about mulching the pansies that you planted?"

Judy says, "What mulching?"

John says, "You told me you were not only going to plant pansies, but mulch them."

John's lawyer comes out and looks at the contract and says, "I don't see any pansy-mulching in here." Will John be able to get Judy to mulch his pansies? No. Because parole evidence says that if this contract represents the final and entire agreement between the parties, all prior and contemporaneous written and oral agreements and discussions are merged into the final document. Since the written contract did not refer to mulching pansies there is no agreement concerning them.

Look at the practical aspects. You go to Orlando to buy a condominium. The salesman says, "This condominium comes with everything you see," and he starts listing all the things that come with it. If he is listing them verbally (and you are not writing them down and getting them in the contract), when you go to settlement and he does not produce—you lose! Whenever a salesman talks about what he is going to do, start writing—and put it in the contract. If when Judy said, "I'll mulch the pansies," John had written that into the contract, it would have become part of the signed contract. She would then have had to mulch the pansies.

Any discussions after the contract is signed are admissible, because in so doing we are now creating a new agreement, and any oral representations which are fraudulently presented can be raised as evidence. If there was a pattern on the part of a salesman who promised everybody a mulching of their pansies in order to induce them to buy when he knew from the start that there would be no mulching done, that is fraudulent.

What we so often hear is, "The salesman told me . . ." If the salesman tells it to you, write it down.

3. Merger vs. Non-Merger

Some contracts merge into others. Some statements merge into others. When parties to a simple contract consent to a subsequent or higher contract, the new contract absorbs the elements of the old contract, and the elements of the old contract are said to merge with the new.

This is important. Using what we know about merger and non-merger, an executory contract merges into the executed contract. If you sign a contract for the purchase of a piece of property and subsequently go to settlement, pay the money, and receive the deed, the deed represents an executed contract. And that is the final contract between the two parties. The previous contract—the sales contract which is the executory contract—dies at that point. There has been a merger of the two contracts.

You have to be aware of merger, but you can also prepare for non-merger. In the case of Judy's and John's mulching transaction, when she said "I'm going to mulch your pansies," John should have written, "The mulching of the pansies will survive this contract." That would be the lay expression; the legal expression is: "The mulching of the pansy provision will not merge with the deed." It is ongoing. If John specifically prevents it from merging, then it does not merge.

To summarize, mulching your pansies is not going to happen unless it is written into the contract. If it is written in the contract but you then sign a final contract, it is still not going to happen unless you provide for a non-merger. If it is not written in the new contract, you cannot come back later and say, "You told me you were going to mulch the pansies." Even if you write it in the executory contract, then go to settlement and the pansies are not mulched you are still out of luck. Therefore get it in writing and make sure that it is a non-merged element.

D. THE ESSENTIAL ELEMENTS OF A CONTRACT

There are five essential elements that must be present in any contract. The three most important are:

(1) There must be two or more parties
(2) There must be mutual assent
(3) There must be consideration

The others, that you should at least be aware of, are:

(4) The two or more parties must have the legal capacity to contract
(5) The object of the contract must be a legal or lawful act

Now that we have an understanding of what contracts are and their essential elements, let's look at the three contractual situations that are most likely to confront you in building and possibly selling your home: the land purchase contract, the construction contract, and the listing agreement.

E. THE LAND PURCHASE CONTRACT

There are several types of real estate contracts. There is a contract for the purchase and sale of improved property, commercial property, rental property, cooperatives, condominiums, and investment properties. When you go to a realtor's office you will usually find a variety of different forms drawn up for different transactions. A contract is binding on both parties but contracts can be written with terms that are more favorable to one party than the other. You can usually obtain blank contracts from the board of realtors in your state. In some states similar (but not identical) contracts are printed in two different colors, one color being more favorable to the seller and the other being more favorable to the purchaser. But the principles that govern the sensible terms of a contract can apply to both sides. The contract we will be mostly concerned with is the Land Purchase Contract.

1. Preliminary Considerations

a. THE PARTIES

Make sure you identify the parties to the contract. If you're the buyer, make sure you identify the seller. If the seller is married make sure that both husband and wife sign. The seller may be a partnership or a corporation. Make sure a general partner signs for a partnership and a corporate officer signs for a corporation. If it is a corporation, make sure there is a corporate resolution that permits the corporation to execute the contract and authorizes the particular officer to sign it.

b. IDENTIFY THE PROPERTY

Make certain that the property is precisely identified. The property should be staked out and surveyed and the sections and parcels correctly identified. Go out to the property yourself, check the stakes, and verify the descriptions. After identifying your property make sure it is correctly identified in your contract. Frequently large subdivisions will contain Sections A, B, C, and perhaps Lots 1 through 25 or more. Occasionally someone will buy Lot 15 in Section A, but by mistake it will read "Lot 15, Section B" in the contract. Not only is there a technical question of ownership, but sometimes in large subdivisions the covenants are different in each section. You should make certain the covenants you are observing are for the section where your property is located.

c. TIME FRAMES

These are periods of time within which you will be expected to perform the contract. In an executory contract there could be two or even three performance dates. One would be the time in which you are to settle on the property. Another would be the time by which you need to complete all of the contingencies to be discussed in the next section of this chapter. There may also be a particular time frame for a financing contingency, another for feasibility contingency, and a time frame within which you have to close. For example, if you bought a piece of property on September 1, you might say, "I want thirty days for an investigation period, I want sixty days to secure my financing, and I want to close thirty days after all these contingencies are met—no later than ninety days from the day of the contract." If you're the buyer, you want to avoid having the expression "time is of the essence" appear in the contract, because it could create a serious problem if you did not meet that deadline.

CONTRACT OF SALE FOR VACANT RECORDED LOT

(For Sale of Unimproved Farmland and Acreage use this Contract with MCAR Form #1306A)

(MCAR Form #1306A is Attached ☐Yes ☐No)

THIS CONTRACT OF SALE, made this _____ day of _____, 19____, by and between the undersigned Seller(s)

(hereinafter referred to as Seller) and

(hereinafter referred to as Purchaser).

1. Deposit has been received from Purchaser with this Contract in the form of _____

in the amount of _____ dollars ($_____).

THE ENTIRE DEPOSIT, RECEIPT OF WHICH IS ACKNOWLEDGED BY AGENT, SHALL BE HELD BY AGENT AND DEPOSITED IN AN ESCROW ACCOUNT IN ACCORDANCE WITH THE MARYLAND REAL ESTATE LICENSE LAW (OR WITH THE APPROPRIATE JURISDICTIONAL LAW) UPON RATIFICATION OF THIS CONTRACT BY BOTH PURCHASER AND SELLER.

WITNESSETH, that for and in consideration of the mutual covenants herein, Seller agrees to sell and Purchaser agrees to buy the property legally described as

Parcel # _____, Tax Plat _____, Lot _____, Block _____, Subdivision _____

and/or Liber _____, Folio _____ also known as (address) _____

_____
(street)   (city)   (zip)

consisting of _____ acres/square feet more or less, located in _____ County, Maryland, including any fencing on

subject property as now installed, upon the following terms of sale:

There are _____ TDRs transferring with this property.

IF THIS CONTRACT IS FOR PROPERTY LOCATED OUTSIDE OF MONTGOMERY COUNTY, A JURISDICTIONAL CLAUSE ADDENDUM SHALL BE ATTACHED, IF APPLICABLE. **Jurisdictional Clause Addendum attached ☐Yes ☐No.**

TOTAL PRICE OF PROPERTY IS _____ dollars ($_____).

If sales price is to be adjusted per survey, an addendum must be attached hereto.

PURCHASER AGREES TO PAY _____ dollars ($_____).

at settlement (by certified, treasurer's or cashier's check) OF WHICH SUM THE DEPOSIT SHALL BE A PART. If the deposit exceeds the down payment, any excess of the deposit shall apply first to settlement costs and the balance shall be refunded to Purchaser at settlement.

2. **FINANCING. a) FIRST TRUST (to be placed or assumed).** Purchaser is to _____ a _____ first deed of trust in lender's usual form

secured by said property of _____ dollars ($_____).

due in _____ (_____) years and bearing interest at the rate of _____ percent (_____ %) per annum,

or the maximum rate prevailing at the time of settlement, payable at approximately

dollars ($_____ _____) per month, PLUS one-twelfth (1/12) of annual taxes and any insurance required by lender.

b) **SECOND TRUST OR SELLER TAKE BACK.** If secondary or Seller financing is applicable, MCAR Addendum Form #1331 is attached ☐Yes ☐No.

c) **FINANCING APPLICATION.** Purchaser placing financing (regardless of type) agrees to make application therefor within ten (10) calendar days of the final ratification of this Contract and agrees to promptly file any supplemental information or papers later requested by the lender and agrees that failure to comply with the terms of this provision shall give Seller the right to declare the deposit forfeited or avail himself of any legal or equitable rights as provided in the paragraph labeled "FORFEITURE OF DEPOSIT/LEGAL REMEDIES."

3. **CONVENTIONAL LOAN.** This Contract is contingent on the ability of Purchaser to secure or receive a firm, written commitment for the herein described conventional financing or lender's approval of assumption, if required, and furnish evidence of commitment or approval to the listing and selling Agents within

_____ (_____) calendar days from the date of final ratification of this Contract, which commitment or approval Purchaser agrees to pursue diligently (see Paragraph 2c hereof). Purchaser reserves the right to increase the cash down payment and/or accept a modified commitment for financing and shall so notify Seller and Agent(s) in writing within the term of this contingency. In the event Purchaser does not obtain the specified financing or increase the cash down payment and/or accept a modified commitment for financing within the specified time period, then this Contract shall be null and void and Purchaser's deposit shall be refunded pursuant to Paragraph 11b of this Contract. By accepting a loan commitment which bears an interest rate or loan amount other than the rate or loan amount designated in Paragraph 2 above, the financing contingency contained herein shall be deemed satisfied and Purchaser hereby waives any rights which Purchaser may have to declare this Contract null and void for failure to obtain acceptable financing. TIME IS OF THE ESSENCE WITH REGARD TO THIS PARAGRAPH.

4. **LOAN FEES. a) Conventional Loan Fee.** If a new loan is to be placed pursuant to this Contract, Purchaser agrees to pay a loan origination and/or discount fee of

_____ percent (_____ %) of the principal sum of ANY CONVENTIONAL LOAN. Seller agrees to pay a loan origination and/or

discount fee of _____ percent (_____ %) of the principal sum of said loan. Lender's fees shall be paid by Purchaser.

Purchaser further agrees to accept any reasonable increase or decrease in said loan origination and/or discount fees where applicable.

b) **Assumption.** If the existing loan is to be assumed, Purchaser agrees to pay any loan assumption fees, charges or expenses required by the lender.

5. a) **EXAMINATION OF TITLE AND COSTS.** Property is to be conveyed in the name of _____

_____ and agrees to pay the settlement charges in connection therewith, tax certificate, conveyancing, notary fees,

PURCHASER HAS THE RIGHT TO SELECT THE TITLE INSURANCE COMPANY, SETTLEMENT OR ESCROW COMPANY, OR TITLE ATTORNEY. Purchaser hereby authorizes the undersigned Agent to order the examination of title and the preparation of all necessary conveyancing papers through _____

incurred if upon examination the title should be found defective and it is not remedied as herein stated. Seller also agrees to pay a reasonable closing fee for services rendered survey where required, lender's fees and recording charges, except those incident to clearing existing encumbrances. Seller hereby agrees to pay any above-mentioned costs to him. Except as hereinafter provided, State and county transfer and recordation taxes shall be paid by _____

b) **COVENANT REVIEW.** Purchaser shall have the right for a period of _____ (_____) days from the date of final ratification of this Contract to obtain and review all covenants, rights of way, easements, conditions and restrictions of record. In the event Purchaser objects to any of these matters of record, he shall have the right to declare this Contract null and void by notifying Seller and Agent in writing of his objection within said time period, in which event this Contract shall become does not qualify for an exemption from the State Agricultural Transfer Tax under Maryland law, a tax of up to five percent (5%) of the consideration for the transfer (in addition to any other transfer tax), excluding the value of improvements, will be assessed for the transfer of agricultural land. If the land is located in Montgomery County, the County may assess an additional County Farmland Transfer Tax which may increase the total of these transfer taxes to a maximum of six percent (6%). These taxes will be paid by

d) REZONING TRANSFER TAX. (Montgomery County only). If the land hereby sold has been rezoned to a more intensive use since July 1, 1971, at the instance of the transferor, transferee, or any other person who had a financial, contractual, or proprietary interest in the property at the time of the application for rezoning, a tax of up to six percent (6%) of the consideration for the transfer, excluding the value of improvements constructed after rezoning, may be assessed for the transfer of the land. This tax would be in lieu of and not in addition to the County Farmland Transfer Tax for this property. This Rezoning Transfer Tax will be paid by

e) TOTAL TAXES. The sum total for the State Agricultural Transfer Tax, the County Farmland Transfer Tax and the Montgomery County Rezoning Transfer Tax will be no more than six percent (6%) of the consideration or the assessed value, if higher.

f) REFUNDS OF TAXES. Any refunds from the real property tax or from any of the above taxes which are generated by payment of these taxes shall inure to the benefit of the payor of said taxes.

6. SETTLEMENT. Seller and Purchaser are required and agree to make full settlement in accordance with the terms hereof on or before the _____ day of _____, 19____, or as soon thereafter as a report of the title and a survey, if required, can be secured if promptly ordered.

7. ADDITIONAL PROVISIONS. SPECIAL PROVISIONS IN THE ATTACHED ADDENDUM, BEARING THE SIGNATURES OF ALL PARTIES CONCERNED, ARE HEREBY MADE A PART OF THIS CONTRACT. ADDENDUM(DA) ATTACHED ☐YES ☐NO.

Front Foot Benefit and water and sewer House Connection Charges to be assumed are

dollars ($ _____) per year as shown on the county tax bill. Seller advises that the cost of deferred transportation related facility charges, if any, are estimated to be _____ dollars ($ _____)

and are to be paid by Seller at settlement. _____

8. AGENCY. Seller recognizes _____ as the Agent(s) negotiating this Contract and agrees to pay such Agent(s) a brokerage fee for services rendered as specified in a separate Listing Agreement. If not previously paid by Seller, the party making settlement is hereby irrevocably authorized and directed to deduct and pay the aforesaid brokerage fee from the proceeds of the sale. However, should settlement fail to occur within the time herein set forth, the Agent(s) shall still be entitled to the brokerage fee herein provided.

Broker or Sales Manager _____

Sales Associate (signature) _____

Sales Associate (print or type) _____ (MCAR #) _____ Broker Code _____ Office Phone Number _____

9. AGREEMENT OF PRINCIPALS. We, the undersigned, hereby ratify, accept and agree to this Contract and acknowledge receipt of a copy hereof. The principals to this Contract mutually agree that it shall be binding upon them, their heirs, executors, administrators, personal representatives, successors and assigns; that the provisions hereof shall survive the execution and delivery of the deed herein stated and shall not be merged therein. This Contract contains the final and entire agreement between the parties hereto, and neither they nor their Agent(s) shall be bound by any terms, conditions, statements, warranties or representations, oral or written, not herein contained. This Contract, any modification, amendment or addendum hereto shall be null, void and unenforceable until Seller and Purchaser have (a) signed or, where appropriate, initialled this Contract and any modification, amendment or addendum and/or (b) transmitted assent through a wired or electronic medium which produces a tangible record of the transmission (such as a telegram, mailgram or datagram) and (c) provided to the other party, in accordance with the paragraph labeled 'NOTICES', the signed or, where appropriate, initialled Contract, modification, amendment or addendum and/or the transmitted assent.

ADDITIONAL PARAGRAPHS NUMBERED 10 THROUGH 28 SET FORTH ON THE REVERSE SIDE HEREOF ARE INCORPORATED HEREIN AND MADE A PART HEREOF AND ALL PARTIES ACKNOWLEDGE THAT THEY HAVE READ SAID PARAGRAPHS.

SELLER _____ PURCHASER _____

SELLER _____ PURCHASER _____

Date of Signature(s) _____ Address of Purchaser _____

Phone: _____ Residence _____ Office _____ _____ City _____ State _____ Zip

SELLER (Print) _____ Phone: _____ Residence _____ Office

SELLER (Print) _____ Date of Signature(s) _____

SELLER OR PURCHASER, WHOEVER PROVIDES FINAL RATIFICATION, IS REQUESTED TO COMPLETE THE FOLLOWING:

Date of Final Ratification: _____ Time of Final Ratification: _____ By: _____ Initials only

This is the **Contract of Sale for Vacant Recorded Lot** recommended by the Montgomery County Association of REALTORS®, Inc.
This Form is the property of the Montgomery County Association of REALTORS®, Inc. and is for use by REALTOR® members only.
Previous edition of this Form may be used until supply is exhausted.
© 1989 Montgomery County Association of REALTORS®, Inc.

MCAR FORM #1306 10/89

10. **SPECIAL NOTICE.** THE AGENT(S) ASSUME NO RESPONSIBILITY FOR THE CONDITION OF THE PROPERTY NOR FOR THE PERFORMANCE OF THIS CONTRACT BY THE PARTIES HERETO. PURCHASER HEREBY WARRANTS AND REPRESENTS UNTO SELLER AND THE REAL ESTATE BROKERS THAT NO AGENT, SERVANT OR EMPLOYEE OF SAID REAL ESTATE BROKERS HAS MADE ANY STATEMENT, REPRESENTATION OR WARRANTY TO THEM REGARDING THE CONDITION OF THE PROPERTY OR ANY PART THEREOF UPON WHICH PURCHASER HAS RELIED AND WHICH IS NOT CONTAINED IN THIS CONTRACT.

11. **FORFEITURE OF DEPOSIT/LEGAL REMEDIES. a)** If Purchaser shall fail to make full settlement, the deposit herein provided for may be forfeited as liquidated damages at the option of Seller, in which event Purchaser shall be relieved from further liability hereunder. If Seller elects not to require forfeiture of the deposit, Seller shall notify Purchaser and Agent(s) in writing within thirty (30) days from the date provided for settlement herein of his election to avail himself of any legal or equitable rights which he may have under this Contract, other than the said forfeiture. In the event that Seller elects not to require forfeiture of the deposit, said deposit shall be retained by Agent holding the same pending resolution of Seller's legal action. In the event of the forfeiture of the deposit, or if Seller shall fail to take any action or fail to pursue any legal or equitable remedies, then and in that event, Seller shall pay the Agent(s) as compensation for services one-half (1/2) of the amount of the deposit, said amount not to exceed the amount of the full brokerage fee. If after a breach by Purchaser, Seller shall release Purchaser from liability hereunder or authorize refund of the deposit, Seller shall pay the Agent(s) as compensation for services one-half (1/2) thereof, said amount not to exceed the amount of the full brokerage fee, but said amount shall not be less than one-half (1/2) of the deposit in the event of a compromise agreement. If the Agent(s) is required to participate in any legal proceeding, either as Plantiff, Defendant or Third Party, Seller agrees to pay reasonable attorney's fees for Agent's own attorney.
b) Except with respect to disbursement of the deposit at settlement hereunder, the deposit and accrued interest, if any, shall be given or returned by escrow agent to any of the parties to this transaction only when an "Agreement of Release" has been ratified by all principals or as directed by a court order. If either Purchaser or Seller refuses to execute a release of the deposit when requested to do so in writing and a court finds that that party should have executed same, that party shall be required to pay the reasonable expenses, including reasonable attorney's fees, incurred by the adverse party in that litigation.

12. **TITLE.** The property, including personal property which conveys hereunder, is sold free of encumbrances, unless otherwise stated herein. Any financing statements will be paid and released by Seller at time of settlement. Title is to be good of record, merchantable and insurable, subject however to the convenants, rights of way, easements, conditions and restrictions of record, if any; otherwise, the deposit is to be returned and sale declared null and void at the option of Purchaser, unless the defects are of such a character that they may be remedied by legal action within a reasonable time. However, Seller and Agent(s) are hereby expressly released from all liability to Purchaser for damages by reason of any defect in the title. In case legal steps are necessary to perfect the title, such action must be taken promptly by Seller at his own expense, whereupon the time herein specified for full settlement by the parties will thereby be extended for the period necessary for such prompt action.

13. **PERFORMANCE.** Settlement is to be made at the office of the Attorney or the Title Company examining the title. Delivery to the Attorney or to the Title Company of the cash payment and settlement costs as herein stated, the executed deed of conveyance and such other papers as required of either party by the terms of this Contract shall be considered good and sufficient tender of performance in accordance with the terms hereof. It is agreed that funds arising out of this transaction at settlement may be used to pay off any existing encumbrances, including interest, as required by the lender(s).

14. **ADJUSTMENTS.** Rents, taxes, water, sewer charges, escrow, insurance and interest on existing encumbrances, if any, and other operating charges are to be adjusted to date of settlement. Taxes, general and special, are to be adjusted according to the certificate of taxes issued by the collector of taxes, except that assessments for improvements completed prior to the date of acceptance hereof, whether assessment therefor has been levied or not, shall be paid by Seller or allowance made therefor at time of settlement. If the property is serviced by the Washington Suburban Sanitary Commission or local government, annual Front Foot Benefit charges and sewer and water House Connection charges of said Commission or local government (which typically appear in the annual county real estate tax bill) are to be adjusted to date of settlement and assumed thereafter by Purchaser. PURCHASER HEREBY ACKNOWLEDGES THAT HE IS ASSUMING ANY OUTSTANDING AND UNPAID FRONT FOOT BENEFIT AND SEWER AND WATER HOUSE CONNECTION CHARGES WHICH WILL BE PAID ANNUALLY.

15. **CONVEYANCE. a)** Seller agrees to execute and deliver a good and sufficient special warranty deed. Purchaser agrees to have the deed of conveyance recorded promptly.
b) Seller or Purchaser, if a corporation, is qualified to do business in the State of Maryland and is a corporation in good standing and is empowered to execute this Contract and is acting pursuant to a duly passed Resolution of its Board of Directors, a copy of which is attached hereto.
c) If either Seller or Purchaser is a general or limited partnership, then such party represents and warrants that it is duly organized and in good standing, is qualified to do business in the State of Maryland, and that any partner executing this Contract on behalf of the partnership is acting pursuant to authority granted to such partner in the Partnership Agreement or pursuant to a duly passed Partnership Resolution, a copy of which is attached hereto.

16. **INSURANCE.** The risk of loss or damage to said property by fire or other casualty until the deed of conveyance is recorded is assumed by Seller.

17. **POSSESSION.** Seller agrees to give possession and occupancy at time of settlement, and in the event he shall fail to do so, he shall become and be thereafter a tenant by sufferance of Purchaser and hereby waives all notice to quit as provided by the laws effective in the state in which the property is located. All notices of violations of orders or requirements noted or issued by any governmental authority or actions in any court on account thereof, against or affecting the property at the date of settlement of this Contract, shall be complied with by Seller, and the property conveyed free thereof.

18. **PROPERTY CONDITION.** At the time of settlement Seller will leave property free and clear of trash and debris. Seller will deliver the property in substantially the same physical condition as of the date of final ratification. In addition to any other specific inspections provided for in this Contract, Purchaser has the privilege of one (1) final inspection of the entire property prior to settlement. Except as expressly contained herein, no other warranties have been made by Seller nor relied upon by Purchaser.

21. **GENERAL/MASTER PLAN.** Purchaser acknowledges that he has been apprised of his rights to review the applicable Master Plan, any adopted amendment to the Plan, and the Wedges and Corridors General Plan for the Bicounty region, including Master Plan maps showing planned land uses, roads, highways, parks and other public facilities affecting the property herein described prior to execution of this Contract. Purchaser further acknowledges that Seller has informed him that amendments affecting the Plan may be pending before the Planning Board or the County Council. Purchaser acknowledges that he has reviewed said applicable plans and adopted amendments thereof prior to executing this Contract or does hereby waive his right to do so. Purchaser also acknowledges that the Agent has advised him of the relative location of any airport or heliport existing within a five (5) mile radius of the property. Purchaser acknowledges that he is aware that the applicable Master Plan or General Plan for Montgomery County is available at the Maryland-National Capital Park and Planning Commission and that at no time did the Agent explain to him the intent or meaning of such a Plan, nor did he rely on any representation made by Agent pertaining to the applicable Master Plan or General Plan.

22. **THE PLAN, GENERAL/MASTER PLANS (CITY OF ROCKVILLE, MARYLAND ONLY).** Purchaser acknowledges that he has been afforded the opportunity to examine the Approved and Adopted Land Use Plan Map portion of The Plan for the City of Rockville and all amendments to said Map (hereinafter referred to as the "Plan"). Purchaser further acknowledges that Seller's real estate Agent has provided said opportunity to examine the Plan by either producing and making available for examination a copy of the Plan or escorting Purchaser to a place where the Plan is available for examination by Purchaser. Purchaser also acknowledges that Seller's real estate Agent has advised him of the relative location of any airport or heliport existing within a five (5) mile radius of the property. Purchaser acknowledges that at no time did the Agent explain to him the intent or meaning of such Plan nor did he rely on any representations made by the Agent(s) pertaining to the applicable Plan. (This paragraph supersedes paragraph 21 hereof only when the property being sold is in the City of Rockville.)

23. **NOTICE AND DISCLOSURE OF AVAILABILITY OF SEWAGE DISPOSAL SYSTEM AND DESIGNATED AREAS. a)** Notice is hereby given, pursuant to the Montgomery County Code, to the prospective Purchaser of the obligation of Seller, or his duly authorized Agent(s), to disclose to Purchaser any information known to Seller as to whether the property is connected to, or has been authorized for connection to, a community sewage system and, if not, whether an individual sewage disposal system has been constructed on the property, whether an individual sewage disposal system has been approved by the County for such property, or whether the property has been disapproved by the County for the installation of an individual sewage disposal system. **b)** Purchaser hereby acknowledges that, prior to entering into this Contract of sale, Seller or his duly authorized Agent(s) provided the above information, as known to Seller or his Agent(s). **c)** If an individual sewage disposal system has been or is to be installed upon this property, and if said property is located in a subdivision, Purchaser acknowledges that he has reviewed the said record plat, including any provisions thereon with regard to areas restricted for the initial and reserve well locations and the individual sewage disposal system, and the restricted area in which construction of the building to be served by the individual sewage disposal system is permitted. **d)** Seller, at his expense, shall furnish to Purchaser, prior to settlement, a written certification by the County Health Authority or recognized private engineer or laboratory stating that, as applicable, well water is potable and that the individual sewage disposal system is in working order.

24. **NOTICE TO PURCHASER AND ALL OTHER PARTIES-GUARANTY FUND (MARYLAND ONLY).** Any person aggrieved in accordance with Article 56A, Section 4-404 of the Maryland Code may be entitled to recover compensation from the Maryland Real Estate Guaranty Fund for his monetary actual loss not to exceed Twenty-Five Thousand Dollars ($25,000). A purchaser is protected by the Guaranty Fund in an amount not exceeding Twenty-Five Thousand Dollars ($25,000).

25. **FINANCING PROVISIONS. a)** In the event that mortgages are used rather than deeds of trust, the word "mortgage" shall be substituted automatically. **b)** If the Contract provides for the assumption of existing trusts, it is understood that the balance of such trusts and the cash down payment are approximate. **c)** Trustees in all deeds of trust are to be named by the parties secured thereby. **d)** Seller shall allow inspections of all of the property and furnish any pertinent information required by Purchaser or his lender in reference to obtaining a loan commitment. **e)** Proceeds of loans acquired pursuant to Paragraph 2 shall be applied to the purchase price.

26. **CREDIT INFORMATION RELEASE.** Purchaser hereby authorizes the Agent to disclose and deliver to Seller or any lender the credit information provided to Agent by Purchaser.

27. **NOTICES.** Notices required to be given to Seller by this Contract shall be in writing and effective as of the date on which such notice is delivered to one of the Agents of Seller named in Paragraph 8 hereof at the principal place of business of said Agent(s). Notice required to be given to Purchaser by this Contract shall also be in writing and effective when either delivered to Purchaser or when mailed to Purchaser's address as shown on page one hereof.

28. **ATTORNEY'S FEES.** Should any litigation be commenced between the parties hereto concerning the property, this Contract, or the rights and duties of either in relation thereto, the party (Purchaser or Seller) prevailing in such litigation shall be entitled, in addition to such other relief granted, to a reasonable sum as and for his attorney's fees in such litigation, to be determined by the Court in such litigation or in a separate action brought for that purpose. If Agent is required to participate in any legal proceedings, either as Plaintiff, Defendant or Third Party, Seller agrees to pay reasonable attorney's fees for Agent's own attorney.

This is the Contract of Sale for Vacant Recorded Lot recommended by the Montgomery County Association of REALTORS®, Inc.
This Form is the property of the Montgomery County Association of REALTORS®, Inc. and is for use by members only.
Previous edition of this Form may be used until supply is exhausted.
© 1989 Montgomery County Association of REALTORS®, Inc.

REALTOR®　　**MCAR FORM #1306**　　10/89

An important note for the seller. If your purchaser includes this contingency provision in the contract, you really don't have much of a contract.

2. Subject To . . .

a. *FINANCING*

If you need financing, make the contract and your purchase subject to financing. Then if you are unable to obtain financing you can usually get out of the contract. You will, of course, have an obligation to try to find financing. The seller will probably help you and perhaps even say, "My brother will finance you." This would be all right if you have written in the contract, "subject to reasonable financing." You should be sure to ask the seller, "What rate will your brother charge?"

Making the contract subject to *market rate financing* is a good rule to follow, because it is more definitive. You can look at a *Wall Street Journal* and arrive at a market rate.

Making the contract subject to financing or even reasonable financing is not sufficient. It is not sufficiently precise and leaves room for dispute. Making the contract subject to market rate financing is more precise and probably offers you adequate protection. But we recommend even more precision, which offers even more protection.

We recommend the following in the contract: "This contract is subject to Purchaser's securing a loan in the amount of $_____." In this blank you insert the largest number possible in order to protect yourself. If you need $100,000, put in $105,000. The Seller may want to put in $95,000. You are trying to get the highest amount while he is trying to get the lowest. Also include "at an interest rate not to exceed _____." If you want 9%, put 8½%. The Seller will probably say 9½%. Also include "for a term not less than _____" [you will probably say forty years and the Seller will say twenty years] and for points not to exceed _____." (Points are explained later in this book.) You are trying to achieve the most favorable scenario. If you are unable to obtain a loan after making reasonable efforts, you would then be relieved of your obligation under the contract and would be entitled to a return of your deposit. With the precision set forth above, a dispute over the financing will be less likely.

The ultimate financing contingency protection for you as the purchaser is to make the purchase subject to the purchaser securing financing acceptable to him in his sole discretion. With this contingency you have total control over whether you want to accept the financing available to you. While this may not seem fair, it is perfectly legal. Remember that although you have the exclusive right to determine if the financing is acceptable to you, you must take affirmative steps to secure financing. If you don't try to secure financing, even if you can reject all that is offered, you are not relieved of your obligation to purchase under the contract.

b. FEASIBILITY

You should have a provision for a period in the contract that will permit you to investigate certain elements of the transaction about which you want to be satisfied before you settle on the property. You could specify an investigation period of thirty or sixty days, or whatever period you negotiate. Once you secure this investigation period and determine a time frame, then the contract is contingent upon the investigation proving to be informative as it relates to certain contingencies. You want to know that the property is properly zoned, which you will investigate. You should check out the covenants of the property and make sure you are satisfied with them and that they are compatible with your concept of a lifestyle. You should find out all you need to know about the neighborhood and the neighbors. And you should make sure that you can get all the requisite permits needed to build your home. Many people will put a provision in the contract that the purchase is subject to securing a health permit. They then find they can get a health permit but cannot get a building permit. So then they might say, "I'll add a health permit and a building permit to the terms." Then they get both of those, only to find they cannot get approval from the architectural review board. Rather than enumerating each permit to the contract you want to say, "This contract is subject to Purchaser being able to secure all requisite permits in order to build his home."

This investigation period is the purchaser's opportunity to check out the property before actually becoming the owner. Use this time wisely. Be thorough in your investigation. Make sure the property suits your needs and that you can build the home you want to live in.

Embodied in both the investigation period and financing is an affirmative obligation on the part of both parties to try to accomplish all prerequisites. If you sign a contract that says you have time to investigate the property, you must investigate it. If you sign a contract and agree to try to get financing at a certain amount at a certain rate you must try to get financing. If you try to get it and fail you are relieved of the commitment. If you don't try to get it, you are not relieved of the contract. You have a contractual obligation to make an effort.

You can include almost any contingency in a contract, although the other party is not obligated to sign it. We knew of one case in which the buyer said he would not close until his grandmother died, and it was written in the contract. If his grandmother did not die, he would not be obligated to buy the house.

Because of the contingency period in a contract during which the buyer is seeing if he can make the deal work and find financing, some sellers include a "kick-out" clause in the contract. A KO clause enables the seller to sell his property sooner without contingencies. A typical kick-out clause would read as follows: "If the Seller notifies Buyer that the Seller has another bona fide buyer that will purchase the property without contingencies, the Buyer will have _____days [some agreed upon time period] to notify Seller that Buyer will settle at the agreed upon settlement date without contingencies."

If the first buyer is unwilling or unable to agree to settle without contingencies, then the seller may void the contract and enter into a non-contingency contract with the second buyer. In this case, the first buyer gives up all rights in the contract and is entitled to a refund of his deposit. A KO clause is frequently used when the buyer has placed a lot of contingencies in the contract or has requested a distant settlement date.

The following is a typical KO clause situation. The buyer agrees to purchase the seller's home contingent upon the sale of the buyer's home. The seller accepts the contract but with a KO provision that if the seller finds another buyer that will settle sooner without contingencies, the first buyer must perform or lose his interest in the contract.

A KO clause is a benefit to the seller. It is negotiable. If you, as the buyer, are not able to perform without contingency, it may be the only way the seller would agree to sell to you.

3. Important Miscellaneous Considerations

a. AFFIRMATIVE OBLIGATION

Both parties have an affirmative obligation to attempt to resolve all the contingencies. If the contract is subject to the purchaser securing financing and the purchaser cannot obtain financing, he is relieved of his obligations provided that he tries to get it and cannot. But if he sits on his hands, or if he goes into the bank and says, "I'm broke," he has not made reasonable attempts to obtain financing.

Quitting a job and then citing unemployment as a reason for not being able to secure a loan is also not permissible. It is not performing positively under the contract. However, losing a job would be a valid reason to withdraw from a contract. As sad as it may seem, when people enter into a contract to purchase a piece of property and subsequently decide to get a divorce before the contract runs out, that does not relieve them from performing under the contract. They are required to take positive, affirmative actions to close on the transaction. Getting a divorce is not a positive, affirmative action.

b. CONDEMNATION
You want the contract to specify that the property is not under condemnation proceedings.

c. ASSIGNABILITY
You want the contract to refer to you, John Doe, "or assigns," which means that you can assign the contract to someone else. The person to whom you assign it must accept it subject to the same conditions. The only way that the assign situation would not work would be if the seller were providing financing to you based on your credit application. In this case, he would not have to assign it to somebody else whose credit might be worse than yours. But if it is a cash deal you can assign the contract to a third party, and he does not have to sign it. This is very important, especially in a real estate contract.

d. CONSIDERATION
If both parties sign a contract for the purchase of a piece of property and the buyer does not leave a deposit, is it a binding contract? Absolutely. It is a binding contract because a promise to buy and a promise to pay are elements of consideration which are the elements of a contract. The deposit does not bind the parties. It is the mutual promises of the parties that make the contract.

e. THE DEPOSIT
The deposit is the money that a purchaser leaves in escrow with either the seller or the realtor to insure his performance. In the event that he does not perform, the deposit can be kept by the seller and/or realtor as liquidated damages. Keep in mind that if you are the seller and the buyer did not put up a deposit it does not mean you do not have a binding contract. What binds the contract is mutual consideration, which is the promise to buy and the promise to sell.

f. NOTICE
It is important that you include provisions for notice in a contract. Notice is how you inform the other party of particular events. Don't get into a situation in which you are the buyer and you don't know where the seller is. You may have to tell him that the contingencies have fallen through. You need to know how to "notice" the seller, and you want the seller to

know how to "notice" you. You don't want to go to settlement and have the seller say, "I couldn't get in touch with you." Make certain that both your address and the seller's address are in the document. Sometimes a contract will have only the seller's name. You should include your address so you will be notified if any problems arise concerning the property or permits.

g. JURISDICTION

You want it specified in the contract under what jurisdiction the contract can be interpreted. If you're buying a piece of property in Maryland, for example, you want it to be interpreted under Maryland laws. You might think this is standard procedure, but it isn t. You need to keep the laws of the contract in a jurisdiction close to home.

3. Default and Its Remedies

a. THE BREACH

When someone does not perform under a contract, this is known as a breach. There is also an anticipatory breach, which means that if one party thinks the other party is not carrying out his part of the contract he doesn't have to wait for the duration of the contract to charge breach. If you have six months to obtain financing but are doing nothing about it, and it is obvious you're doing nothing about it, the other party does not have to wait six months to sue you. If, after a reasonable length of time, he can prove you're doing nothing to obtain financing, he can sue.

When a purchase contract is breached there is a breaching party and there is an aggrieved party, and it is important to know what these terms mean. The breaching party is the one who does not perform. The aggrieved party is the one who stands to suffer thereby.

b. REMEDIES

If a real estate contract is breached the remedies available to each party are different. If a buyer agrees to pay $100,000 for a piece of property and the seller agrees to sell it and the buyer subsequently breaches, the seller is entitled to receive damages for the breach. The seller would be entitled to the lost profits he would have made on this particular transaction because the seller has a right to be made whole. At this particular point, the seller is obligated to mitigate the damages. The seller can try, using reasonable efforts, to sell that piece of property. If he sells it for more than the price the buyer agreed to pay, then he has suffered no damages. But if he sells it for less than that—say $90,000—the buyer is liable for the amount between the contract price and the price at which it could be sold within a reasonable time frame and in a reasonable market—in this case $10,000.

It is different if you're the seller and you breach. If you're the seller and you agree to sell that property for $50,000, and then decide "I'm not going to sell it for $50,000" (probably because it is worth more than $50,000), the remedy available to the buyer is called specific performance. He can sue and require you to perform specifically under that contract and to deed the property to him for the sum of $50,000. The reason for the two different approaches to remedying a default is that the law says it would be very difficult to determine damages if the seller breaches. The property is unique, you have a contract for it, and you're entitled to it.

When you sign a contract be prepared for the consequences of a breach, depending on whether you are the buyer or the seller. If you're the seller and there is a realtor involved, you have an obligation to the realtor if that realtor produces a buyer who is ready, willing, and able to buy. As the seller you need to be ready to make arrangements to satisfy the interests of the realtor in the event there is a breach by either you or the buyer.

If you're the buyer there should be a provision in the contract that in the event of a breach by the seller (or in the event of litigation), the seller must pay the buyer's attorney's fees. If you're the seller, you may want to include a provision that in the event of a breach by the buyer, the buyer will pay the seller's attorney's fees. These are permissible additions to a contract, and you should keep in mind which status you occupy and protect yourself accordingly.

c. LIMIT YOUR LIABILITY

If you're a buyer and you execute a contract for the purchase of a piece of property and you subsequently breach, you are liable to the seller for his damages. One way to limit your liability is to state in the contract that, as buyer, in the event that you do not perform you agree to forfeit your deposit and that the forfeiture of the deposit will relieve you from any further liability under that contract. If you don't have a provision in the contract that permits you to forfeit the deposit and relieve you of any further liability and you breach, then you may be responsible for the seller's entire damages—not just the loss of your deposit.

d. LIQUIDATED DAMAGES

When you sign a contract you frequently put up a deposit. That deposit will often equal an amount that has been provided for liquidated damages. This means that in the event of breach rather than having to litigate to determine the damages, you have predetermined them. If there is a breach by the buyer he will forfeit the amount of money specified as liquidated damages.

This eliminates the cost of litigation. So frequently the deposit on a contract is made to equal liquidated damages. Be careful, though. Some contracts will say that if you, the buyer, breach, you are liable not only for the deposit but for any other legal or equitable remedy that the seller may have under the law. As a buyer, you want your contract to read that the buyer leaves "a deposit in the amount of _____." You then state the amount to be used as liquidated damages in the event the buyer does not perform under the contract. Then if the buyer does not perform under the contract he forfeits his deposit and is relieved of any further liability under this contract.

Of course, if you didn't perform under the contract because the contingencies were not met then the issue of liquidated damages is not relevant, and accordingly your full deposit would be refunded.

F. THE CONSTRUCTION CONTRACT

Study the contract that follows. In the thirteenth century, there was a philosopher named William of Occam who said, "What happens if we have two competing theories and they both have the same intellectual strength? They both make sense. Which theory is the better?" The answer was the simpler of the two, and thus the postulate, "Occam's razor," which is often cited in game and risk theory and economics. This contract is a classic example of Occam's razor. You could use this contract to do ninety percent of the contracting for your home.

1. The Players

In drawing up construction contracts it is important to know who the players are and to keep their distinctions clearly in mind. Under the law a general contractor is anyone who contracts directly with the owner. For example; you hire a builder, John Jones, to build your home. He's usually known as a general contractor. But let us assume you also hire Tom Smith, an electrician, to do the electrical work. He's also a general contractor because he contracted directly with the owner. If you're the owner and are also managing the building of your own home then you are both the owner and general contractor. In this case the electrician is also a general contractor because he contracted directly with the owner of the house. But if John Jones had hired him instead of you then Smith, the electrician, would be a subcontractor.

In the practical sense a general contractor is somebody who manages the building of your home and hires the other trades. But from a legal standpoint a general contractor is anyone who contracts directly with the owner. From a legal standpoint it is important to make a distinction between the general contractor and the subcontractor. In building your own

Specifications, Proposal & Contract

Owner_____   Date _____

_____   Project Address _____

_____   _____

Phone (Day) _____ (Eve.) _____   _____

Contractor _____   Type of Work _____

_____   _____

_____   _____

Phone (Day) _____ (Eve.) _____   Job # _____

SCHEDULE START DATE _____   COMPLETION DATE _____

Description and Specifications of the work to be performed:

_____

_____

_____

_____

_____

_____

_____

_____

_____

_____

_____

_____

_____

_____

Allowances _____

_____

Contractor agrees to furnish labor and material to complete the above described work for the sum of _____

_____

Payment to be made as follows: _____

_____

All work is to be performed and completed in a workmanlike manner in substantial accordance with the plans and specifications. Contractor agrees to work expeditiously to complete the project.

Owner agrees to carry a builders risk insurance policy. Contractor agrees to maintain both workmans compensation insurance and general building insurance and to provide Owner with evidence of same.

Other Terms and Conditions: _____

_____

Witness the following Signatures and Seals:

Owner _____   Contractor _____

Date _____   Date _____

home most of the people with whom you will be contracting will be general contractors, regardless of their job description.

In addition to the *Owner* and *General Contractor,* the other players are:

The Subcontractors or Mechanics: The subcontractor is a worker or mechanic who contracts with the general contractor. A mechanic is an old English term, somebody who performs a trade—a plumber, an electrician, an insulation mechanic, a trim carpenter, or framing carpenter. All trade people are mechanics. A *mechanic's lien* is the right of somebody who has performed the work on your house to claim an interest in that house for nonpayment for services that he has performed.

Materialmen: The materialmen supply the material. It could be lumber or it could be drywall. A mechanic can be not only a mechanic but also a materialman. So there can be a mechanic's lien or a materialman's lien; the effect is the same.

2. The Essence of the Contract

In the contract you need to list the parties to the contract and to specify exactly what is to be done. And you need to specify the cost and payment schedule. The more specific you are in every category, the better. If a contractor is going to install the heating, ventilation, and air conditioning, he should identify the quantity of material and the brand name or that what he will furnish will be comparable to a brand name. If he says, "Carrier or comparable" for the air conditioner, for example, that's a legal term. But it has to be something that is comparable to a Carrier. If you want only a Carrier, scratch out the words "or comparable" and make sure you both initial the change.

When the contractor signs the contract it is then an offer to furnish and install. He will specify a time frame (generally ten days or thirty days) within which it can be accepted. If you don't accept it within that time frame, the offer expires. If you do sign it within that time frame, it is an executory contract.

All material is to be guaranteed as specified; all work is to be completed in a "workmanlike" manner. That is a legal term, meaning it has to conform to the industry standards in your particular area. Any alteration or deviation from these specifications involving extra cost will be executed only upon written order. You can alter it orally if it is less than $500, but we recommend that all changes, no matter how small, be in writing.

The contract also makes reference to workmen's compensation insurance. Workmen's compensation insurance is probably not sufficient coverage. You contractor should have general liability insurance as well. Ask him for a certificate from his insurance carrier for both types of coverage. The greater the amount of work he is to perform the more important the insurance is.

This contract does not specify that the work must "conform to code" (the local regulations). You can insist on that but you will usually have to write it into the contract. And be sure to include provision for attorney's fees in the event that you have to sue.

Some people like to specify that in the event of a dispute it will be turned over to arbitration. We do not recommend this. Studies have shown that in arbitration there is an effort to compromise on the problem rather than to solve it on the merits. And no matter how the arbitration agreement is drafted, in all probability the party who loses arbitration is going to sue anyway. So arbitration just delays the inevitable.

Look for the elements included in our contract in whatever form your contractor uses. If he doesn't have a form, we recommend this one; it can be bought at any stationery store. Don't make your agreement on an envelope while standing on your property, because you could omit something important.

3. The Payment Schedule

The payment schedule needs to be well thought out. We can illustrate this point with a story. A home improvement contractor comes to your neighborhood and agrees to build five family rooms for five different homeowners. He gets out of his car, looks at your house, and starts walking around, still looking, he already knows that it's going to cost him $10,000 to do the job, and he'll probably mark it up a hundred percent, charging you $20,000. But he's taking plenty of time in order to make his estimate look studied and solid. He quotes the same price to all five homeowners, but they all decide on different payment schedules.

The first, Mr. Stingy, says, "Tell you what I'm going to do. I'm going to give you $2,000 now and $18,000 upon completion." Down the street is Mr. Quickdeal. He says, "I'll give you $10,000 now and $10,000 when you complete the job." The next guy is Mr. Reasonable. He says, "I'm going to pay you $5,000 now, $5,000 when it is 25% done, $5,000 when it's 50% done, and $5,000 upon completion." The next guy is Mr. Smart. He says, "I'm going to give you $1,000 now and then I'm going to give you $5,000, $5,000, $5,000 and $4,000. That adds up to $20,000." The last guy is Mr. Justright. He agrees to put down $2,000, make three payments of $5,334 as the work progresses, and pay $2,000 when the work is completed.

The contractor signs up all five homebuilders. Which house does he start on first? You can be sure he does not start on Mr. Quickdeal's house. Quickdeal would be the last one he starts because he already has $10,000 without doing any work. And he's not going to start with Mr. Stingy's home first because he'll receive only $2,000 up front and must then complete the job before he receives another cent.

The point to remember is that you have to position yourself somewhere between too stingy and too generous with your contractor. Both the Mr. Reasonable and Mr. Smart deals are all right but we prefer the arrangement made by the last one, Mr. Justright. It does three things: (1) it brings and keeps people on the job, because as they work they receive instantaneous reward, (2) it never gives the contractor more than ten percent of the contract price at a time—any more would be providing capital for the contractor (which should be unnecessary), and any less would make him think you are stingy, and (3) at the end, it does not hold back a large sum and give him the feeling he will never be paid. Holding back at least ten percent assures that he will finish the job. This is the 10/80/10 rule, which is the best payment schedule—never more or less than ten percent to begin with; never much more or less than ten percent at the end; and the middle eighty percent should entail as many payments as practicable. This will keep the contractor's morale high. It is very important to establish a good rapport with your contractors. Frequent payments for work performed will do more to get your home built quickly than any other incentive or contract provision we know. Overpaying or underpaying will increase the likelihood that your contractor will avoid your job site. He's going to go somewhere where he gets paid when he does some work. A fair payment program is the most effective way we know to keep the contractor(s) on the job.

4. Licensed, Bonded, and Insured

You will find that almost all contractors are licensed, bonded and insured. But you should check to make certain that they are properly certified. The contractor should have two licenses—a trade license and a business license which authorize him to perform his trade in your county. A trade license is granted him by the authority that regulates his trade, and it is very important. A business license is a permit to operate a business in that county. From your standpoint a business license is not as important as the license for his trade. You should make certain that the plumbers' association in your area certifies that he knows how to plumb.

You also need to know for how much and for what he is bonded. A bond for $1,000 isn't going to help you much if he can't perform. The word "bonded" on the side of a truck does not really mean much. However, if you pay him for his work and materials on a reasonable draw schedule, the you don't have to worry about whether a contractor is bonded. If he isn't overpaid and doesn't finish the work your loss will be the administrative effort to find another contractor rather than a financial loss for paying for work not done.

5. Quantum Meruit

The legal term *quantum meruit* means that a person is entitled to be paid a reasonable amount for the work that he has done even if he doesn't complete the entire contract. It makes no difference whether you fire him or he quits. If a plumber contracts to plumb your house for $5,000 then quits (or gets fired for cause) after he's done half of the work, do you owe him money? The answer is yes. You owe him for that portion of the contract he has completed. But, if he has done half the work does he get $2,500? Probably not. You only owe him the difference between what it will cost you to have the job completed and $5,000. When someone quits halfway through a job the new person will probably charge more than half the total amount of money to complete the job.

6. Mitigation of Damages

When a person breaches a contract the aggrieved party has an obligation to mitigate (minimize) the damages that result from that breach. In the above example of the plumber who quit after completing half of the work you must make reasonable efforts to find someone to complete the job for the best price possible. (Remember, the first plumber gets the difference in the contract price and the completion price.) You cannot take advantage of the first plumber by agreeing to pay $4,000 to complete the job if the second plumber would have done it for less. Also, if the first plumber has left his material or tools out in the weather you must make a reasonable effort to keep them from being damaged. Even if you're upset with him you are legally responsible for mitigating (minimizing) his damages.

g. THE LISTING CONTRACT

As we have said, most people sell their homes within eight to ten years. Hence, sooner or later, you'll probably need to know something about listing contracts. A listing contract is an agreement between you and somebody who is going to sell your property.

1. Types of Listing Agreements

There are essentially four types of listing agreements. The first one is called an open listing.

a. OPEN LISTING

An open listing is one in which you contract with a realtor or broker and permit him to sell your property. If he sells your property he gets a commission. But the open listing also permits two other arrangements: (1) You can sell your property yourself, and (2) another broker or realtor can also sell your property. You could enter into an open listing with three or four brokers or realtors.

An open listing specifies that anybody who has an agreement with you can sell your property. Of course the selling agent must be a realtor or a broker under the law. You cannot just have anybody sell your property who wants to do so, except for yourself. You can sell your own property whether or not you are a realtor or broker.

b. EXCLUSIVE LISTING

An exclusive listing gives only one realtor or broker the authority to sell your house. But it also permits you to sell your house. There is a big difference in this agreement and the third example below.

c. EXCLUSIVE AUTHORIZATION

The exclusive authorization to sell is an agreement wherein the broker or realtor gets the commission from the sale of your house regardless of whether he sells it, you sell it, or John Doe sells it. What usually happens with exclusive authorization to sell is that the broker or realtor who is given the exclusive authorization then places your house in what realtors call "multiple listing." This means that any recognized real estate agency in the area can sell the house. The commission is then split between the selling agency and the agency that was given the exclusive authorization. If that agency sells the house it takes the full commission. The difference between exclusive agency and exclusive authorization is that in the former you can sell your home without paying a commission.

d. NET LISTING

A net listing is when a broker or real estate agent agrees to sell your property and give you a net amount from the proceeds. If you want to sell your house for $100,000 net and you net list your property with a realtor for $100,000, that gives the realtor the opportunity to sell it for $120,000, or $101,000, or any other amount. The realtor or broker then keeps the amount in excess of the $100,000. A net listing is frequently frowned upon by realtor associations and is actually prohibited in some states. Obviously, it is not recommended unless you want to encourage a massive effort on the part of your realtor or broker to dispose of your property.

2. Termination

Listing agreements are usually made for a specific time period or until the house is sold. The seller specifies how long the listing will last—thirty, sixty, ninety, one hundred-twenty days, or even a year. A listing agreement is a binding contract as long as the realtor is putting forth an effort to sell your house. However, if you're not satisfied most realtors will let you out of a listing agreement after a reasonable length of time although they are not obligated to do so.

Most listing agreements state that not only will the broker be paid a commission if the house is sold during the time frame of the listing agreement, but that if the broker has introduced somebody to your house, and that person subsequently buys your house within a secondary predetermined time frame the broker still gets a commission. The reason for this is obvious. If someone comes to your house on the eighty-ninth day a ninety-day listing and says, "I'm thinking about buying," you could say, "Don't buy today! Tomorrow my listing agreement is up and you can come back and buy it from me for the listed price minus the realtor's commission." The listing agreement protects the realtor in that situation.

H. IMPORTANT CONTRACT EXPRESSIONS AND TERMINOLOGY

In most contracts you'll find legal expressions and terminology. Some are very important, and you should be aware of what they mean.

1. Penalty Clause

An agreement between two or more parties where a party agrees to pay a sum in excess of the actual damages in the event of nonperformance. Often confused with a liquidated damages provision, a penalty clause is unenforceable. For example, if a general contractor agrees to pay an owner $1,000 a day for every day of late delivery and the actual damages to the owner are less than $1,000, the clause is unenforceable.

2. Liquidated Damages

A damage amount predetermined by the parties in the event of a breach. An example of liquidated damages would be one in which the purchaser forfeits his deposit when he fails to settle on the contract. Another example would be a provision whereby the contractor agrees to pay the owner the sum of $100 per day if the home is not completed by a predetermined date. If the $100 is a good estimate of the potential damages then it is a permissible provision. If the figure was $1,000 and was out of line with the potential damages, then it would be a penalty clause and therefore unenforceable.

Whether you call the provision a liquidated damages clause or a penalty clause is immaterial. The effect is the controlling factor. The breaching party can only be held liable for the actual damages. Damages in excess of actual damages are penalties and will not be enforced. Don't attempt to spur performance by including damage sums in excess of your potential damages, as it will only backfire on you.

3. Time is of the Essence

This is a legal term and it means, among other things, that if you're supposed to settle on a property at 12:00 noon on April 21st and you show up at 1:00 P.M., you're out of luck! What if your husband or wife died on the way? Is that a defense? No. It doesn't make any difference what happens, if it says 12:00 noon on April 21st, and you don't close by then you lose your deposit, you lose the property, and you lose the deal. If you're the buyer, you may want to delete this phrase because it puts you under an ironclad deadline to perform. If you're a seller, of course, then you may want to include it.

4. As Is

As we shall discuss later, there are certain kinds of warranties, expressed, implied, and statutory involved in a contract. However, when you purchase something with the wording "as is," then all warranties, expressed or implied, are waived. It is a legal term and whether it relates to a new house, a used house, a vehicle, or a toaster, it eliminates any warranties.

5. Caveat Emptor

A Latin phrase that translates as "let the buyer beware." There may be situations in both real estate and contract law when you should be aware that you're not covered by all the protections you might expect. In this case you are usually forewarned by the term "caveat emptor."

6. Substantial

You will see the word "substantial" in contracts, and it's not a bad word. The building of a house is an imperfect process. Don't expect that every 2×4 is going to be perfectly straight or every shingle perfectly flat. When the contract says that the builder or plumber or electrician will perform in substantial compliance to the plans and specifications, don't be too concerned. It merely means that building the house is not a perfect process and the house will be built in substantial compliance. For example, if the plans call for a house to be 44 feet by 28 feet in exterior dimensions and the house is 44 feet by 28 feet ¼ inch, that is substantial compliance. If the contract said it was to be built in substantial compliance with the plans and specifications you wouldn't be able to avoid paying the contractor because of that quarter-inch difference.

7. Workmanlike

This means the work will conform to the standards of the industry in your area. If a builder or subcontractor says, "Don't worry about my work because the county's going to inspect it," don't accept that. The county does not inspect for quality control, and it won't involve itself in a dispute between you and a contractor over the quality of the work. The county will look at the work from the point of view of compliance with building code regulations (for example, how deep the footings are) but not the workmanship.

8. Reasonable

"Reasonable" means that people of average intelligence within the community would perceive the facts to be fair and equitable to all parties.

9. Consequential Damages

This is an important term which is best explained by citing an actual case, *Hadley v. Baxendale*. A large wheel broke causing a mill to shut down. The mill owner called a transportation company to take the wheel back to the manufacturer for repair. The transportation company picked up the wheel, hauled it off to the manufacturer, had it repaired, hauled it back to the depot, and then lost the wheel! The mill owner then sued the transportation company for losing the wheel.

The cost of the wheel was around $1,000. The mill owner also sued the transportation company for the lost profit associated with all the flour it could have made while that wheel was gone.

Losing the wheel is called direct damages. You lost your wheel. It's worth $1,000. You're entitled to $1,000. What about the lost profits?

The lost profits are called consequential damages. The mill owner could not win consequential damages because the transportation company couldn't foresee that the absence of the wheel meant the mill owner could not grind the flour. It could have logically assumed that the mill owner had another wheel available.

The mill owner could have told the transportation company, "Take this wheel to be fixed, because I'm losing money every day that it's not fixed," and the transportation company might have said, "I may lose that wheel. If I lose the wheel, I'm not only going to have to pay for the loss of the wheel but for the loss of the profits, so I'm going to charge you more for transporting it." If the mill owner had told the transportation company about the potential consequential damages and the transportation company agreed to take on the task, including in its price the risk that the transportation company was concerned about in lost profits, the mill owner would have been entitled to the consequential damages.

In most contracts concerning the building of a home, in the event of a breach (specifically late delivery or late performance) the owner would be entitled to direct damages; but it's highly unlikely that he would be able to obtain any consequential damages such as rent or lost profit.

10. Mitigate

"Mitigate" means to soften the negative impact of a breach. In contract law the non-breaching or aggrieved party (the party that suffers under a breached contract) has an affirmative obligation to mitigate his damages. If a person has sold a piece of property and the buyer backs out the seller of the property has an affirmative obligation to take reasonable measures in order to try to sell the property to someone else. A landlord who has a tenant who walks out on a lease must take affirmative action to try to rent the property and at least mitigate or lessen the impact of the breach.

11. Fixture

A fixture is an item affixed to a house or a piece of property. If it is permanently affixed it becomes real estate. Examples of fixtures in a house would be a radiator, wall-to-wall carpeting, plumbing fixtures, or a hot water heater. These things could be detached but only with a great deal of difficulty. Since they are permanently affixed and an integral part of the house, they are termed fixtures. As fixtures they are real estate, and as real estate they are governed by a different type of law than personalty.

12. Personalty

"Personalty" are items such as furniture and clothing. It should be clearly understood that when you buy a house from someone you don't get the furniture and you don't get the clothes. But some things are not so clearly understood. For example, draperies that can be easily removed are likely to be considered personalty and not fixtures. Venetian blinds, on the other hand, have been interpreted to be fixtures because they are permanently affixed.

When you enter into a contract to buy real property you take title to all the real estate but you don't take title to the personalty. You take title to real estate by virtue of the deed. You take title to personalty by a bill of sale.

13. A Bill of Sale

The instrument under which you sell and purchase and subsequently take title to personalty, as distinct from real estate.

14. Insurable Interest

If you own a house you have an insurable interest in that house because if the house burns down you have a loss. If you have a spouse you have an insurable interest in that person because the loss of that person would be detrimental to your physical, emotional, and financial well-being. You have an insurable interest in a business partner provided the loss of that business partner would have an adverse impact on your income. You must have an insurable interest in real estate, personalty, or an individual in order to obtain insurance on that realty, personalty, or individual.

3 **Designing Your Home**

Before you can start building your new home you'll need a concept and a set of blueprints which will tell the carpenters, plumbers, electricians, wallboard men, and others exactly what they have to do.

There are essentially three ways to design and/or select a house plan: hire an architect, purchase a plan from a plan service, or work with a builder who already has plans. In all of these cases you can modify the plans to meet your particular needs.

People frequently ask, "Do I need an architect to design a home for me?" It depends. An architect will create a unique home just for you. And if you want a unique home just for you that might be the way to go. But it's expensive and may not be necessary.

There are thousands of plans from plan services that are available at very reasonable prices (generally a few hundred dollars). Most builders will provide you with plans at no cost.

In all of the above cases you can make changes to personalize the home.

Although you'll find that most custom home builders and architects have put a great deal of thought into the designs of their homes, here are a few tips for realizing maximum value from your custom home:

A. THINK RESALE

One of the main reasons for building a custom house is to create your own personal home and environment, even at the expense of conventional features. However, you still have to obtain financing, and the bank or savings and loan company will want your house to be attractive to other potential buyers. More than likely, too, you'll want to sell your home sometime in the future. So the wise builder will "think resale" from the beginning. This is a check to make sure you are not overdoing your project or creating a white elephant.

1. Unique Home Means a Limited Market

A unique home means a limited market. A one-bedroom house may be quaint and perfectly adequate for you and your spouse but it would be almost impossible to obtain financing on a one-bedroom house or to sell it. In varying degrees this is also true for all house designs with features that depart radically from the conventional house with two to five bedrooms, living, family, and dining rooms, kitchen, and two to three bathrooms. The conventional design is what most lenders prefer and what most house-hunters are looking for. Unusual features that are attractively incorporated into your home may make it more exciting and perhaps make it sell a little faster. But as a rule the conventional size and design of a house should conform to broad-based tastes. You will have a much easier time obtaining a loan and/or selling your home when they do.

A good crosscheck for your project is a willingness of a lender to finance the home you are building. Even if you're uncommonly qualified financially, your home may be too unique. If so, think twice before proceeding.

2. The Common Driveway

Almost invariably in custom home construction you will find friends or even new neighbors getting together and building a common driveway. Common driveways can be very dramatic and impressive as well as practical. In some cases they might even be essential for one of the homebuilders. A typical common driveway situation might be the one shown in the illustration.

The common driveway is fine as long as you and your neighbor are getting along. But the day might come when you and your neighbor have a falling out. Then you could face a situation that is not only difficult to live with, but could conceivably hinder the future sale of your house. If possible, avoid the common driveway or any other features that you share with a neighbor.

3. The Neighborhood—and the Neighbors

In Chapter 1 we discussed the importance of the neighborhood and your neighbors. Their relationship to the value of your land is obvious. Often

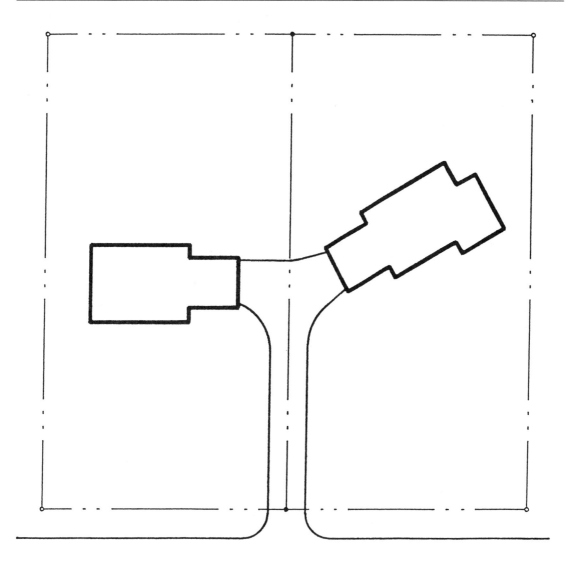

the first thing people look for is the neighborhood. Then they try to find a piece of land or a house in that neighborhood. Most people know what makes a neighborhood attractive to them, but it's important to give some thought to which way the neighborhood seems to be going. What will it be like in five or ten years? It's not always easy to predict a neighborhood's future direction but you should at least give it some thought.

Don't overlook your individual next-door neighbors. A very difficult neighbor can not only be a pain in the neck to live next to, but his presence might even affect the future sale of your house.

4. What to Emphasize to Maximize Resale Value

Back in the 1960s Secretary of Defense Robert McNamara helped popularize the phrase "cost effective" in analyzing national defense programs and weapon systems. How much "bang for the buck" did you get from each weapons system? Was the system worth it in terms of the cost?

The same kind of analysis is often applied to housing construction. What is the cost of a certain kind of window, or door, or closet, and what is the benefit? Does the benefit justify the extra cost? If it does, it is cost effective. The more the benefit exceeds the cost, the more cost effective it is.

When you are landscaping, picking materials, and planning special rooms and areas, you will have some difficult cost-benefit decisions to make. Eventually you will be confronted with a number of options, especially when it comes to material. More often than not the cost-effective approach is the best one and will perhaps make your decisions easier. Here are a number of things to consider in designing and building your home. These suggestions are based not only on our experience but on research by the American Homebuilding Institute of Fairfax, Virginia.

a. *WET AREAS*

By wet areas we mean bathrooms and the kitchen. They should be light, bright—and tiled. Special bathtubs and Jacuzzis have become very popular and can be considered very cost effective.

b. *CLOSETS*

They should be large. Everybody knows this, but they still don't make them large enough. Closets should also be creative. A good architect knows that there are ways to make closets different from the conventional walk-in closets set along one side of the wall. You can put them in a corner and save space. And don't forget broom closets, coat closets, and linen closets. The more closets, the better.

c. *LANDSCAPING*

Landscaping has a cost benefit curve that is very easy to plot, as is shown in the illustration.

What the cost benefit curve shows is that up to a certain point you receive substantial benefit for the money spent on landscaping. But as you increase your spending the benefits begin to flatten out, and at the end they start to dip. Spending $5,000 on landscaping will probably provide optimum benefit. If you spend $10,000, however, you may end up adding only $8,000 worth of value to your property. And a house that is extremely overlandscaped may actually begin to reduce in value.

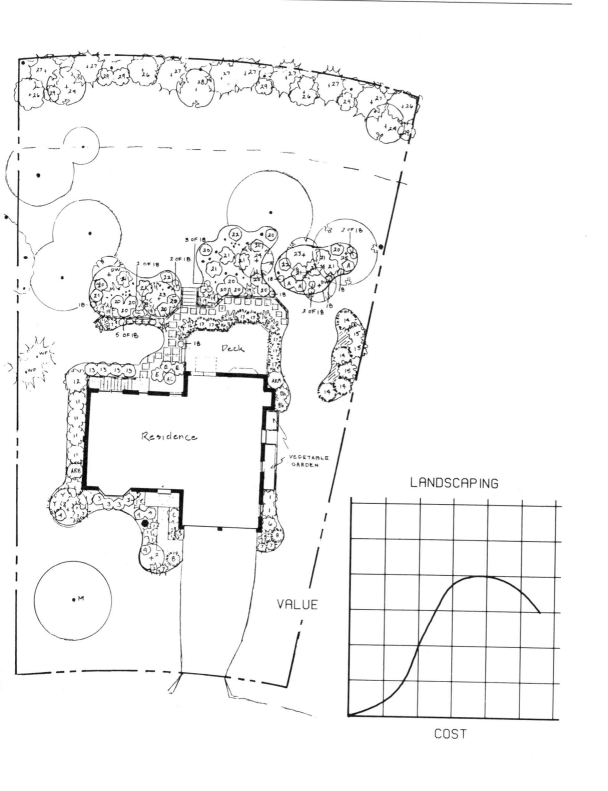

LANDSCAPING

VALUE

COST

d. THE SIX-INCH EXTERIOR WALL

We feel that, by the twenty-first century, any conventionally framed home that does not have six-inch external walls will be functionally obsolete. (Block and stucco homes offer adequate insulation, strength, and mass, and are not subject to the changes that are going on with frame homes.) A home with six-inch exterior walls offers the following: (1) better insulation, (2) more strength, and (3) more mass. Irrespective of the geographic area, we strongly recommend a six-inch exterior wall. The cost is minimal and the benefit is significant. Don't be left behind in design, strength, and insulation. Design and build your frame home with six-inch walls.

We also recommend that you specify that the stud spacing be kept at sixteen inches or under rather than increasing the spacing to twenty-four inches. The more studs and the thicker the wall, the better built your home will be.

e. QUALITY—INSIDE AND OUT

(i) Framing Lumber

When you home is finished you don't see the framing lumber. But what you don't see is important. We recommend that you specify #2 or better lumber. If you don't specify the grade, your builder may use a utility-grade lumber which is not what you want. A custom home is not only your design, but your specifications. The quality of your framing lumber is as important as your plumbing fixtures.

(ii) Fiberglass Shingles

Fiberglass shingles are new and they are extremely popular in the home-building industry because they last longer and they have a Class A fire rating. (Asphalt shingles in contrast have a Class C fire rating.) Some people think the fire ratings for roof shingles are not important, since fires generally start in the house below and burn upward, but this is not always true. Many fires start on the roof and burn downward.

Fiberglass shingles require more care to install and they're difficult to work with during the winter because they are brittle. Asphalt shingles, on the other hand, are difficult to install in the summer because they have a tendency to squish out with the heat.

Slate and tile roofs also add major drama to a house, but from a cost standpoint they should only be considered for fairly expensive homes. And cedar shake shingles are cherished items even though they may not be cost effective on moderately priced homes.

(iii) Name-Brand Windows

Using name-brand windows will definitely be a good investment. Four out of five people have heard of Andersen windows. Less than five percent of the people on the street have heard of any other window. Andersen builds extremely good windows. There are other companies that build windows that are as good as Andersen's, but since they don't spend millions of dollars a year promoting their windows, the name-brand identification for their windows is not very high. People do not demand them. An Anderson window can help sell a home.

(iv) Doors

A door is a statement about your house. A metal door is an energy saver while a wooden door is an energy loser, although the loss of energy through the door is really negligible. So a decision on a front door should be made on aesthetic grounds.

A wooden door—especially an exotic wooden door made of mahogany, oak, or the like—has a tendency to warp, so when you're building your custom home and plan to install a wood door, consider installing a cheap temporary door at the beginning of construction. Wait until the house is finished before you install the final door because if it isn't properly sealed it will warp immediately. Many times the builder installs a great door and tells the customer to seal it. The owner procrastinates. Three days later there's a rainstorm and the door warps.

There is a door system that is new on the market which includes two doors. The jamb and inexpensive door is installed for use during the construction period. When the house is ready for occupancy, the final, more expensive door is installed and leveled.

(v) Subfloors

The subflooring in your home is structural. It's the support between the floor joists and your finished flooring. Although it cannot be seen it's very important. Unfortunately, subflooring is an area where builders often cut costs.

You've seen the movies in which a person buying a used car kicks the tires as part of the inspection process. The same is true for the flooring system of your home. Invariably, people bounce up and down on the floor to check for springiness or squeaks and, invariably, both are found. And there's a tendency for the buying public to accept this condition as something that is standard rather than to expect something more. You can prevent both springiness and squeaking by proper design and proper construction techniques. First, the joists should be thick and narrowly spaced. The subfloor plywood should be a ¾″ thickness (⅝″ is the industry standard). It should be both nailed and glued to the floor. Finally, the spaces between the joists should be secured with metal cross bridging, as shown

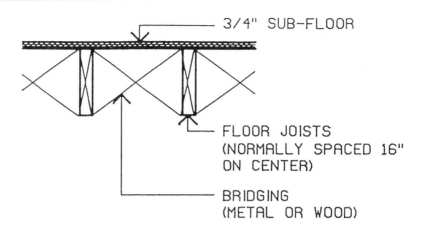

in the illustration. These simple procedures will prevent springiness and squeaks.

(vi) Window and Door Installation

It's important that windows and doors be properly installed. The headers are the horizontal framing over windows and doors. They're supported by vertical jacks as shown in the illustration. The headers and jacks form a frame around your doors and windows and keep the pressure off them, especially as your house settles over the years. Ask your builder for bigger headers, and double headers over your windows and doors. Ask him to double the window and door jacks as well. This will give you more stability in the windows and doors and protect them against warping. It should cost less than ten percent more per window and door and will be money well spent.

(vii) Decks

Decks clearly add value to your house and are very cost effective. The same could be said about gazebos and porches. Homeowners like the appearance of these horizontal extensions, and we all want to be outside when the weather is nice.

(viii) Foyers

Foyers are popular now. They went out of style in the 1960s and '70s, but they are now a much-desired item.

(ix) Vestibule

An even more popular installation. Vestibules are enclosed foyers and you see them often in northern and New England states. They are not only exciting from a design point of view but they are very energy efficient, and their use is spreading to other parts of the country.

(x) Garages

Garages are very important. The standard garage ranges from 20′ × 20′ to 24′ × 24′ in size. If you increase the size beyond 24 feet the additional cost is minimal and the benefit immense. An oversize garage is extremely cost effective. Garage storage space is as alluring as the oversize garage. Work and storage areas in a garage are also items that are extremely attractive to prospective buyers. Working space in the garage is usually much more desirable than working space in the basement.

(xi) Basements

Basements are extremely cost effective. It costs approximately $10 a square foot to add a basement but the value to the buyer is well beyond that figure. If your land makes it possible, a walk-out basement adds even more value. And if you build a walk-out basement, you should add a nice sliding-glass

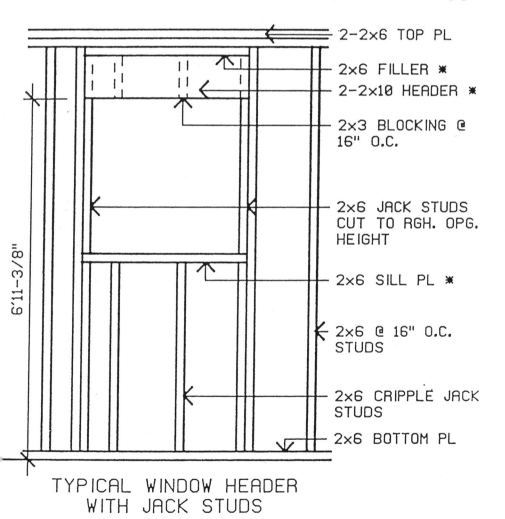

TYPICAL WINDOW HEADER
WITH JACK STUDS

door and some windows. A walk-out basement will cost about the same as an in-ground basement, but because of the aesthetics will be even more cost effective. A frequently overlooked specification on the basement is its ceiling height. Depending on your region, a basement may have interior heights of 7'0" to 7'8". We strongly recommend that your basement height be 8'0". It will cost only a few dollars more to add another course of block or some more concrete and forms if you have a poured foundation. And the extra height in the basement is a welcome asset.

A basement foundation system is best suited to a northern or cold climate region. Cold-climate foundations have to be placed deep in the soil to avoid heaving as the ground freezes, and it is logical to add the basement space by excavating all the earth within the walls and pouring a slab. Having an enclosed basement provides a warmer first floor even without heating the basement. However, the basement is a logical location for the heating plant, which can be designed to include the basement space in its heating load. Negative characteristics usually associated with basement space include dampness, the cold concrete floor, lack of natural light and ventilation, and the size and location of the stairwell. Construction techniques used to build a basement today can almost eliminate all of these negatives. A basement is still one of the most economical ways to increase the usable space in a house.

Crawlspaces can be used in all but the northernmost states, where the practical problems of sufficiently insulating floors, water pipes, and heat ducts curtail their use. Crawlspaces are best used in moderate to warm climates where minimum insulation is required. The floor of the crawlspace can be a concrete slab but is usually level ground covered with a vapor barrier of polyethylene. The principle advantage of a slab in a crawlspace is the ability to use the area for storage. Also, the heating and cooling equipment is frequently located in the crawlspace, and a slab base will eliminate the need to suspend the equipment from the floor system. Crawlspace foundations are less expensive (by about 50 percent) than basements but do not provide the usable space. Crawlspaces must be properly ventilated to avoid condensation, which can cause rotting of the wood floor structure.

Slab-on-grade foundations are used primarily in warm climates. Although the footings can be placed low enough for colder climates and the perimeter of the slab can be insulated to reduce heat loss, conventionally constructed concrete slab floors are cold when outside temperatures remain at or below freezing for extended time periods. Specially designed slab floors can be excellent for solar-heated spaces because of their ability to retain heat. For cold climates, slab floors can be heated mechanically with air ducts or water pipes. This provides a very comfortable floor and living space, but the cost of the heating equipment is much higher than more conventional forced-air heating systems. Standard slab-on-grade foundations are usually less expensive than crawlspace foundations.

A pier or piling foundation system is best used in coastal areas (velocity-wind zones), in flood plain areas, on property with soils that will not accept conventional spread footings, on property with steep or irregular grade conditions, and in warm climates. The advantages of this type of foundation are that wind (and water) passes easily around it and the structure it supports and that it requires a minimum of bearing points in the soil. Since it totally exposes the floor of the structure, it is not suited for cold climates. The cost of pier or piling foundations can vary widely because of the soil conditions, terrain, type of support, number of supports, structural strength requirement, and location of the building site. For a normal (beach front) wood piling foundation the cost is equivalent to a crawlspace foundation. Each of these foundations can be cost effective if you consider your own needs, the climate, and the specific requirements of your property when making your decision.

(xii) *Cathedral Ceilings and High Walls*

People talk about the loss of energy of cathedral ceilings, but they love them anyway. In fact, the loss of energy is minimal. Cathedral ceilings come in many sizes and forms and can add a very dramatic appearance to a house.

Another dramatic feature is when the entire ceiling height exceeds the usual eight-foot height. A room with a nine-foot or a ten-foot ceiling, especially on the first floor, creates a very striking effect. Unfortunately, the additional cost of building a nine-foot or ten-foot wall is not insignificant because most building modes come in eight-foot lengths. They can still be cost effective. If you are building a large, imposing house, we recommend that you do something a little bit different and make your ceilings a foot or two higher on the first level. You can reduce the height to eight feet on the second level.

(xiii) *Bookshelves*

Built-in bookshelves are pleasing to the eye, functional, dignified, and cost effective.

(xiv) *Multiple-Level Rooms*

Walk-ups and sunken rooms are cost effective in both contemporary and traditional homes.

(xv) *Angular Glass Gables*

Any glass gable is cost effective, but the new angular glass gables and those with the little curvilinear caps on them are much more effective than square glass gables. Round windows are also popular—even half-round windows. These are very dramatic and very cost effective, although they're not cheap.

(xvi) Fireplaces

Although fireplaces are definitely not energy efficient, people love them, and they're still cost effective. Masonry fireplaces are real energy losers, but they add aesthetic appeal and considerable value to your home. A metal or free-standing fireplace is less costly and more energy efficient, but it doesn't add as much value as a masonry fireplace.

(xvii) Special-Purpose Rooms

Special-purpose rooms—a sewing room, library, workshop, exercise room, or a music room—are, within limits, very cost effective. However, as we shall see, there are some special-purpose rooms that can have a negative impact, or are cost ineffective.

(xviii) Great Rooms

Extra-large living rooms that often contain the family area are becoming very popular, and even appear to be replacing the conventional living room. This may be only a present-day fad, however, and the conventional living room may well stage a comeback. Where possible (depending on your life-style) we recommend a large, spacious great room with lots of features and a small dignified living room for your less frequent formal needs.

(xix) Washer and Dryer Connection

They are essential, and the main concern today is not whether to include them but where to locate them. They are many places they can go—utility room, basement, kitchen, or master bedroom. It's a subjective decision, and it's all but impossible to predict where the next owner of your house might want the washer and dryer. A special laundry room is the ideal place, of course.

(xx) Dishwasher

This is an absolutely critical component of your house—if not for you, for the resale of your home.

(xxi) Microwave Oven

Clearly cost effective, a microwave oven will probably cost you around $300 and add $400 to $500 in value in your house. Since it is an essential element, it should be designed into your home rather than added as an afterthought.

(xxii) Deadbolt Locks

People are turning more and more to deadbolt locks because they offer more security. In fact, some insurance companies offer reduced premiums if they are installed on all outside doors.

(xxiii) Name-Brand Hardware

The type of hardware installed in a house is important. A recognizable name brand adds value and status to your home. Start with a nice front-door handle.

5. What to De-emphasize to Maximize Resale Value

Just as the previous items are known to be cost effective and add value to your home, there are a number of additions that have proven cost ineffective. The decision on whether to install them should be made based on your own personal desires and interests. If you have a personal reason to add them, go ahead—but not because you think it will add value to the house's resale value. If you very much want a swimming pool and you think it will be continually used, by all means have one built, but not because you think it will add value to your house and you ought to have one. A future buyer may not want a swimming pool—and may not want the cost or nuisance of the upkeep.

a. FINISHED BASEMENTS

A basement itself is cost effective, but finishing it may not be. For every dollar you spend on finishing your basement you'll be adding less than a dollar in value to your home—not only in terms of the appraisal, but in the resale price. The basement itself is cost effective but finishing it may be cost ineffective.

b. SWIMMING POOLS

Swimming pools, as popular as they are, are not cost effective. They will cost much more than the value they add to your house.

c. LIVING ROOM

A separate living room is not cost effective in the small to mid-size house. You're better off to emphasize the great room.

d. ELECTRONIC GADGETRY

Generally speaking, electronic gadgets are not cost effective. There are some exceptions, however. If you can install a burglar system for around $1500, it would be cost effective. Above that price it probably would not be. Equipment that is complicated to use and expensive to repair has a limited appeal to most buyers.

e. SPECIAL ROOMS

As we've seen, some special rooms are very popular and will increase the value of your home. But research has proved there are others that do not. A poolroom, video room, astronomy room, and special workshops fall into the latter category.

B. ENERGY EFFICIENCY

1. Home Design and Insulation

For most of us the most significant energy cost is heating. The important point to remember about heat loss is that approximately eighty percent of your heat loss is through the ceiling. Accordingly, it is imperative that you design your roof system for a maximum level of insulation for your climate.

The remaining fifteen percent of heat loss is through your walls, and of this fifteen percent, eighty-five percent is through and around your windows and doors. Accordingly we recommend the following, regardless of your climate:

1. All exterior walls should be 2″ × 6″ (not only for insulation, but for mass and strength).
2. All windows should be insulated. (In colder climates they should have storm windows and in sunny climates they should be tinted.)
3. The openings around all windows and doors should be sealed to prevent air infiltration. This can be done by either stuffing or injecting fiberglass or (preferably) insulation foam in the areas around windows, doors, and any areas where the wall has been cut (such as an electrical outlet).

The last five percent is lost through the floor areas. Insulate your floor and include a sill cushion between your foundation and home. A sill cushion is simply a gasket that seals the area between the masonry foundation and the wood structure. Sill cushions, although frequently forgotten, are essential features of a well-built home.

2. Mechanical

Heating and cooling systems vary depending on your climate. Don't be too experimental in this area. Rely on what your region has been using effectively. In many areas of the country the heat pump, which provides both heating and cooling, will be the most economical and energy-efficient system. However, the heat pump can only heat effectively when temperatures are above zero degrees. It heats more efficiently in the 32 to 60 degree range. Accordingly, if you're in an area where evening temperatures are consistently below 32 degrees you should have a backup system. A backup heating system (such as oil or gas) will operate when the heat pump is inefficient and the heat pump will take over when temperatures are in its efficient range.

Another drawback to the heat pump is that the heat coming out of your register is not the warm heat you're used to feeling. It warms your home and keeps it comfortable, but you won't feel the warm heat that comes out of the more conventional system.

3. Solar Homes

All homes use solar energy to some extent. There are essentially two types of solar homes—active and passive. For the house as a whole, all studies show that an active solar house is not cost effective because the capital cost to put in the system is never recovered from the savings generated by the system itself. An active system also requires a great deal of maintenance. It uses moving parts to generate heat or cooling and in today's environment they're not cost effective. However, although active systems are not cost effective overall, a solar hot water heater is apparently more cost effective than an electric hot water heater, but not as effective as oil or gas.

On the other hand, the passive solar homes, which do not require any mechanical assistance to generate heat or to cool, are cost effective. There are two types of passive solar homes—direct gain and envelope homes. The direct gain system is best illustrated by a story we read about a young woman who survived a plane crash in the Sierra Madre Mountains. The young woman, who was the only survivor, was seriously injured. She managed to struggle out of the plane, and found herself up on a rocky mountainside. She could see the valley below but no one could see her. During the day it was warm, but at night it was bitterly cold. She realized that she would not make it through another frigid night. She found gasoline on the plane, poured it on the rocks, and lit it. The heat was absorbed by the rocks. At night she lay on the rocks and was able to keep warm until she was found.

This is exactly what happens when you have a patio; the tile or natural dirt retains the heat of the sun. The same principle is used for a passive, direct-gain solar energy home, and it's very effective. The system is constructed so that the heat from the sun during the day is retained and then radiates out during the night. In the right geographic areas, such a system is very cost effective.

The other type of passive solar home is based on the envelope concept. The envelope solar home is a house within a house.

In the right areas, the envelope home can be cost effective, although not as cost effective as the direct gain, because there are some significant initial costs associated with building an envelope home.

C. THE SQUARE FOOT MYTH

When people start to think about building a home they immediately begin to calculate how much a particular home would cost. Unfortunately, there is a time-honored industry yardstick, frequently used, called the square-foot method. The SFM determines the cost per square foot of home. When you have a square-foot cost it enables you to make quick comparisons not only between two different-sized homes, but between builders who quote you SFM figures to build them.

The SFM of cost comparisons should be avoided! You don't purchase cars by the pound and you should not purchase homes by the square foot. If all homes and all builders were the same, the theory would work. But all homes and builders are not the same, and the SFM does you and them both a disservice.

There are several reasons why SFM cost analysis is inappropriate:

(1) Home styles vary. The SFM is not helpful in comparing different styles—a contemporary to a traditional home, a brick home to a wood home.

(2) Home design. The SFM cannot be accurate in comparing a two-story home to a one-story home.

(3) Specifications. The SFM really falls apart when you have the same home design and style but different specifications. The SFM gives an inaccurate comparison of cost/value when one home has brick, hardwood floors, solid cherry cabinets, and Kohler plumbing fixtures while the other home has Masonite siding, carpeted floors, laminate cabinets, and basic white plumbing fixtures. Does the SFM include site work, well, septic, landscaping, basements, garages, fireplaces? Some quotes include all of those, some quotes include some of those, and some include none. So, for comparative analysis, it's not helpful.

(4) Home size. The SFM is totally inadequate when one attempts to compare a 1,000 SF home to a 5,000 SF home.

(5) Room definition. If the same size home has one floor plan with

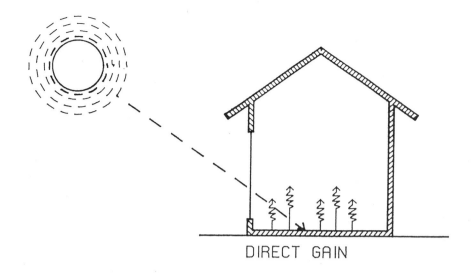

DIRECT GAIN

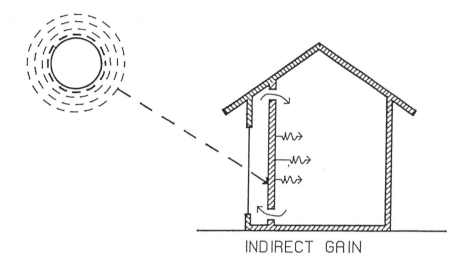

INDIRECT GAIN

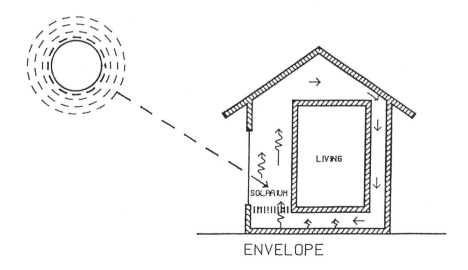

ENVELOPE

four bathrooms and another plan with two bathrooms, the SFM would create a skewed comparison.

(6) Marketing considerations. Builders know that some people rely on SFM value analysis and they tailor the SFM bid to meet your requirements. In order to match your square-foot cost number they may knowingly or unknowingly downgrade the specifications to accommodate your square-foot cost budget.

In summary, don't comparison-shop homes or builders based on SFM. It may have been an effective tool in the days of Abraham Lincoln and log cabins, but it's totally inappropriate and in fact dangerously misleading, in today's housing market.

These are a few tips for getting the most value out of the money you spend on the various features of your house and determining which features you include or exclude. No single item may have much of an effect one way or the other, but taken together they can make a significant difference in the value of your final product.

4 The Essentials of Surveying

Surveying may be the most boring aspect of building a house, but it is also one of the most critical tasks you will have to undertake. It is important that you understand the essentials of surveying. If you do, you will save yourself numerous headaches and a considerable amount of time and money.

A. THE REALITY OF REQUIREMENT

The first thing you have to deal with in surveying is what we call "the reality of requirement." Have you ever gone to a bank to apply for a loan? If you have a good income and plenty of credit-worthiness, you no doubt feel there is no reason why you should not be able to obtain a loan. The well-dressed loan officer sits behind his desk, with a booklet in his hand and asks you questions. You're applying for a $1,000 loan and he asks:

"What's your name?" You tell him.

He asks, "Middle name?" And you tell him.

"How do you spell it?" "What's your spouse's name?" "Middle name?" "Where are you from?" "Where were you two years ago?" "Where were you five years ago?" "Where were you born?"

And you sit there, asking yourself; "What is this guy doing? He's an idiot."

But worse, not only is he asking these dumb questions, but you're giving him the answers. Why?

Because he has the money—and that is the reality of requirement. The identical scenario exists in surveying. The more intellectual you are the more difficulty you are likely to have with the surveying process. But you must comply with the surveying needs of the people building your home or you won't get far.

Who needs your survey? Almost everyone involved in the building process, including some of the tradesmen, will need your survey. The excavator needs to know where to clear the house site. The builder needs to know where to place the house. The county and architectural review board need to know that the house conforms to their setbacks. The title company that insures your property wants a survey, and so does the lender. Solution: Get a survey, get it early, and make it complete.

B. ROMAN LAW

The second thing to remember when you consider surveying is that you must think three-dimensionally. If you think only two-dimensionally you'll run into considerable trouble and worry. A bird's-eye view means looking down; a human view means looking straight ahead. If you combine the two you end up with three dimensions.

This story involves two ancient Romans whom we will call Tully and Cicero, which will illustrate three-dimensional thinking. They were good friends and both had good incomes. They grew tired of the traffic in Rome with all the chariots and dust, and one day they decided to go outside of Rome and buy a large tract of land and build a home.

Tully said "Well, I can't afford to buy a five-acre tract by myself."

Cicero said, "I can't either. We'll buy ten acres together and subdivide it."

So they went out, bought ten acres, and built very simple homes.

Their land was as flat as a pancake and very barren. At this point Tully and Cicero were very good friends. They built separate houses on their own subdivided property, shared a chariot together, went out on the town together, and worked on their chariots on weekends. Then one day Tully said: "It's awful barren out here." So he planted an olive tree which took a number of years to grow large and produce fruit.

Tully planted this olive tree on his side of the property line. But over the years the olive tree grew larger and its branches extended over Cicero's property as well. However, as the tree grew and prospered the relationship between the two Romans deteriorated. Ultimately they were feuding continually and Cicero decided he would do anything he could to complicate Tully's life. One day he said, "Your olive tree's branches are on my property!" And he sued Tully for civil trespass. They went to trial. Who won?

Cicero won. The ruling was that the overhanging branch constituted trespass, and that ruling is applicable today in most western societies.

This is why we want you to think three-dimensionally. Consider not only where you're going to position your house on your property, but how high up the roof is going to be, how deep in the ground the basement is going to be, and whether any portion of the house is going to hang over a setback or property line.

C. KNOW THE PLAYERS

In order to begin the survey process, it's essential that you know who the players are. There are five of them: the owner, the builder, the surveyor, the county (or city), and the lender. They all want to know the dimensions of your land and where you're going to build your home.

D. THREE IMPORTANT SURVEYS

1. The Proposed House Location Survey

The first people to be involved in the survey will be you and your spouse, if you have one. You should go out to your property yourselves, walk around, and decide where you're going to position your home. Using the two-dimensional bird's-eye approach, look down on the land and see where you want your house to be. Using the human-eye view, decide how far down in the ground you want your basement. Then imagine how high up the roof line will be. Imagine views and approaches. What will it over-hang? Then select your site and place your markers for the location of your house.

The next person to be involved will be the builder. Bring the builder out and show him where you've positioned your home. The builder may walk over to the edge of the cliff where you've positioned your house and tell you, "I can save you $5,000 if you move your house back about ten feet."

Your spouse says, "Why? I like the view."

The builder says: "We don't have bulldozers with wings and we can't fly around out there."

"All right, how about here?"

"No, not back there, because this big rock is going to cost you $5,000 to move. Let's move it over here away from the rock and the edge." So you, your spouse, and your builder position your house. "Why didn't I bring the builder out to begin with?" you ask. If you had come out first with your builder he would have immediately said, "Let's build it right here," at the easiest spot on your property on which to build a house. You should pick your spot first and then have your builder tell you what, if anything, is wrong with your position. If he convinces you that your position will cause problems or undue expense then you should work together to reach a compromise position.

The next person to be involved is the surveyor. You need a survey to make sure the house is positioned within the property line, within the county setbacks and covenant restrictions, and complies with all easements (such as a power line) and is not on your septic field or the like.

The surveyor, finding that your house is now properly positioned, will prepare a document called a Proposed House Location Survey.

The Proposed House Location Survey is sometimes referred to as the site plan or grading plan. We prefer proposed house location survey because the term "proposed" denotes that something is planned, something is to be done. The PHLS is a critical component for the following parties:

(1) The Builder. It tells the Builder where to position your home on your property. If properly done it tells him not only left and right placement but depth and height.

(2) The County. In most jurisdictions the County will require the PHLS to accompany your plans and specifications as part of the building permit application. The permit department will review both documents for accuracy and code and zoning compliance before a permit will be issued.

(3) The Title Company. Before a lender will make you a loan he will require certification of clear title and survey compliances. Accordingly, the title company will require the PHLS (just as the county does) to review it for title compliance. The PHLS must show that your project complies with zoning, covenants, and easements before the title company will insure your project. Without the title insurance you cannot obtain funding.

The PHLS as outlined above is the first survey document you'll need, and you'll need it before your project can begin. Once all of the approvals have been obtained, the surveyor goes to your property and physically stakes the home site following the placement on the PHLS.

2. The Wall Check Survey

The next step includes the owner, the builder (who might also be you), and the mason. The mason comes out, lays your foundation, pours the concrete, and puts in your foundation walls, carefully following the stakes which show where the foundation is to be built. When the foundation is

PARCEL "A"

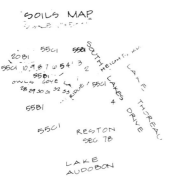

SOILS MAP

SOILS LEGEND

20B1 – MEADOWVILLE SILT LOAM, MIXED ALLUVIAL LAND
55B1 – GLENELG SILT LOAM, UNDULATING PHASE
55C1 – GLENELG SILT LOAM, ROLLING PHASE

N 27° 21' 59" W 78.00'
EX LAT INV. 333.91

LOT 5
12,161 ☐

CLEARING & GRADING LIMITS

DECK 352.50
GREAT ROOM = 352.76
BF. 344.00
CO INV = 341.50
ENTRY = 354.29
#2108
DECK 353.80

CLEARING & GRADING LIMITS

CLEARING & GRADING LIMITS

GARAGE = 357.50

RAISED WALKWAY

EX WATER A = 59.65'
EX DE-1
EX. CROCK FC.T

R = 300.00

OWLS COVE LANE
(50' WIDE)

S 02° 28' 01" W
N 09° 03' 19" E
170.91'
190.23'

LOT 4
LOT 6

GRADING PLAN
LOT 5
BLOCK 1 - SECTION 78
RESTON
CENTREVILLE DISTRICT
FAIRFAX COUNTY VIRGINIA

SCALE: 1" = 20'   DATE: APRIL 9, 1986

CHARLES R. JOHNSON
CONSULTING CIVIL ENGINEERS & LAND SURVEYORS

SUITE 21

11250 ROGER BACON DRIVE
RESTON, VIRGINIA 22090
(703) 471-9370   CERTIFIED CORRECT

LOT # 144

COMMONWEALTH OF VIRGINIA
CHARLES R. JOHNSON
CERTIFICATE No. 1144
CERTIFIED LAND SURVEYOR

completed the mason wants his money. So you go to your lender for your next draw. The lender will ask for your wall check survey.

The wall check survey is simply a confirmation by your surveyor that your foundation was actually built where the house location survey said it would be built. We have seen grown men and women go berserk when told by the lender that they needed another survey to confirm the first survey. Before the lender will advance more money, however, he must be absolutely certain that the house is being built where it was supposed to have been built, and the only way he can have proof of that is with another survey of the actual existing foundation. The reason he is insistent on this survey is that if there has been some mistake and the house is built in violation of your setback requirements, there is no recourse. The county or your property owner's association could insist that the entire structure be torn down even if you have completely finished your house. Obviously the lender cannot take a chance on that—and you shouldn't either.

Now is the time to check your project for compliance before you proceed with expensive improvement. The cost of the wall check is minimal.

3. The Final Survey

When your home is completed most lenders (there are only rare exceptions to this requirement) will require a final house location survey before they will advance the final draw on your loan or grant you your permanent loan. The requirement generally surprises and frustrates the homeowner. "Why do I need this now?" the often irate owner asks. "You already checked the foundation. Do you think I moved the house?"

The lender's answer is, "No, we don't think you moved your house, but we want to be certain that the eaves on the roof, the chimneys, and the decks are also in compliance." Remember the story of Tully and Cicero? We've seen houses where foundations comply with setback requirements but whose roof lines do not. The penalty for failure to comply is removal, not compromise.

So there are three major surveying requirements: the proposed house location survey, the wall check survey, and the final house location survey.

Know these surveys well, as you'll generally need all three.

E. THE SURVEYING SEQUENCE

Let's take a broader look at the surveying sequence. In all probability the first survey at which you will look is your boundary survey—a two-dimensional portrayal of your property as the bird sees it. It tells you a great deal, but not everything, about your property.

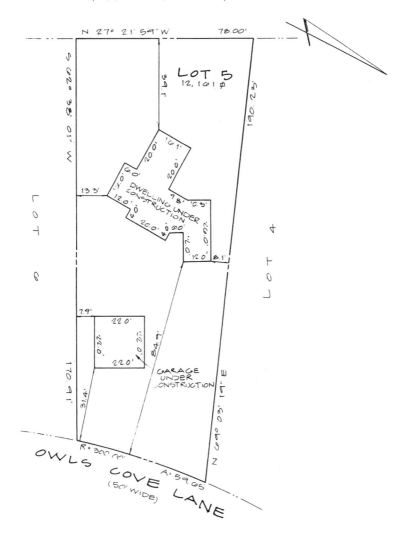

PARCEL "A"

N 27° 21' 59" W   78.00'

LOT 5
12,101 ф

S 62° 38' 01" W

190.23'

LOT 6

LOT 4

DWELLING UNDER CONSTRUCTION

13.3'

GARAGE UNDER CONSTRUCTION

170.41'

R = 300.00'   A = 59.05

OWLS COVE LANE
(50' WIDE)

WALL CHECK
LOT 5
BLOCK 1
SECTION 78

RESTON

CENTREVILLE DISTRICT
FAIRFAX COUNTY   VIRGINIA
SCALE 1" = 20'   DATE: AUGUST 19, 1980

CHARLES R. JOHNSON
CONSULTING CIVIL ENGINEERS & LAND SURVEYORS

11480 SUNSET HILLS ROAD
RESTON, VIRGINIA   22091
(703) 471-9570

CERTIFIED CORRECT _Charles R. Johnson_
LS # 1144

COMMONWEALTH OF VIRGINIA
CHARLES R JOHNSON
CERTIFICATE No.
1144
CERTIFIED LAND SURVEYOR

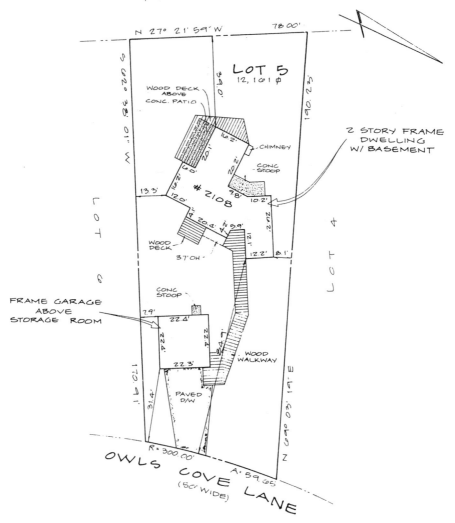

PARCEL "A"

N 27° 21' 59" W   78.00'

LOT 5
12,101 ϕ

WOOD DECK
ABOVE
CONC. PATIO

CHIMNEY

CONC.
STOOP

2108

2 STORY FRAME
DWELLING
W/ BASEMENT

13.3'

12.0'

WOOD
DECK

37' OH

20.4'

5.9'

12.1'

12.2'   8.1'

FRAME GARAGE
ABOVE
STORAGE ROOM

CONC.
STOOP

7.9'

22.4'

22.4'

22.4'

22.3'

WOOD
WALKWAY

PAVED
D/W

31.4'

R = 300.00'   A = 59.65

OWLS COVE LANE
(50' WIDE)

BOUNDARY &
HOUSE LOCATION SURVEY
LOT 5
BLOCK 1
SECTION 78
RESTON
CENTREVILLE DISTRICT
FAIRFAX COUNTY   VIRGINIA
SCALE 1" = 20'   DATE 11 JUNE 1987

CHARLES R. JOHNSON
CONSULTING CIVIL ENGINEERS & LAND SURVEYORS

11480 SUNSET HILLS ROAD
RESTON, VIRGINIA 22091
(703) 471-9370

CERTIFIED CORRECT Charles R. Johnson
LS # 1144

COMMONWEALTH OF VIRGINIA
CHARLES R JOHNSON
CERTIFICATE No.
1144
CERTIFIED LAND SURVEYOR

1. The Topographical Survey

Of equal importance is the topographical survey—a three-dimensional portrayal of your property. It depicts not only boundaries but the high and low portions of your property. Knowing the high and low points are important for the proper placement of your home for appearance, access, views, drainage, and building cost. A topo map is an important element in the building of your home.

With a topo map you, your builder, and your architect can select the best location for your home. With instruction from you, your architect, and the builder, the surveyor can now lay out your home on the property not only within the boundary and setback requirements (a two-dimensional condition) but also positioned vertically to fit the contours of your property. This placement will determine the precise three-dimensional position of your home.

2. Elevation and Cut and Fill Data

The placement of your home within your property or setback lines is relatively easy. The surveyor simply puts stakes in the ground. But how does a surveyor tell the builder the vertical positioning of your home? He does it with elevation readings or marks. All elevations are keyed to sea level. If the first floor of your house is to be 208 feet above sea level, the surveyor will state that on his survey (site plan). He will then provide the builder with a bench mark, which is a measurement above sea level. For example, the surveyor will nail a stake on a tree at a point located at 210 feet above sea level and mark it "210′." The builder then knows that the mark in the tree, the bench mark, is 210 feet above sea level. He can now adjust the foundation of the house so that the first floor will be at 208′.

The surveyor should also supply cut and fill data. Based on the surveyor's knowledge of your land from the topo map, he can tell the builder where to cut the land down and where to fill in the land so your home will be exactly where you want it.

The surveyor provides information (stake, bench marks, cut and fill data) to place the home not only within the property lines but within the property vertical space.

F. MISCELLANEOUS CONSIDERATIONS

If you have a wooded lot there is a two-step process for the surveyor to follow: The first step is for the surveyor to stake the area that is to be cleared for the driveway, septic tank, and homesite. Make sure that he identifies which trees are to be left and which are to come down. After the site is cleared, he returns to stake the actual homesite, including bench marks. Do not stake the actual homesite prior to clearing because the clearing process may destroy the stakes, requiring the field survey to be done again.

Make sure your surveyor stakes your house location with offset stakes. Offset stakes are set some predetermined distance away from the actual corner (generally 10 feet). This gives the builder room to work around the corner without knocking over the stakes.

G. TECHNICAL TERMS

In discussing your survey with the surveyor, the architect, or the builder, certain technical terms (some of which we have already discussed) will come up and you should be familiar with them:

Bench mark—This is a reference point the surveyor will put on or near your property to indicate height. It will usually be a stake nailed on a tree with a figure written on it. The builder uses this bench mark to determine how high the house will be or how high the first floor will be. In some areas you have to build to a certain height because of the floodplain, but usually you will be concerned about the height of your house because of a view that your want to preserve.

Elevation—This is the term used in discussing the height of your house, the roof, or one of the floors.

Topo lines—These are the lines on your survey that show the height or elevation of the land.

Accretion—Let's say you have a piece of property on which the boundary line is determined by a stream or a river. Gradually the river or stream changes course, thereby changing the boundary lines of the property. The law recognizes that property lines determined by streams or rivers can be changed, and sometimes there are lawsuits and a court has to decide what is fair. Accretion is the term used for land legally gained by a stream changing course gradually over time.

Reliction—This is the term for land legally lost by a stream gradually changing course.

Avulsion—When a sudden jump of a stream or a river causes the sudden removal of soil from the land of one owner to another—usually caused by a storm or flood—this is called avulsion. There is no change of property ownership or boundary lines when avulsion has occurred.

5 Financing

A. THE CONSTRUCTION/PERMANENT LOAN

When you're ready to build your custom home, it's essential that you properly identify the loan for which you are looking. What you need is a construction/permanent loan (often referred to as a C/P loan). A construction/permanent loan is actually two loans in one. The construction loan is the loan you will need to take the home from concept to occupancy. As you build the home the lender will advance funds to you at various stages of your progress. It is a short-term loan, payable when your house is completed. After you finish building the house and are ready to move into it, you'll need a permanent loan, which will pay off the construction loan and enable you to live in the home while you make affordable monthly payments. Before your project can be properly completed you'll need both a construction loan and a permanent loan. So when you go shopping for a loan keep in mind that you're looking for both of them.

B. SELECTING A CONSTRUCTION/PERMANENT LOAN

There are at least thirteen elements to consider in finding a construction/permanent loan. They all have varying degrees of importance and you'll have to determine which are most important to you.

In looking for a construction/permanent loan, it's a good idea to consult a checklist of criteria that will help you select the loan that's best for your project. Some of these elements are more essential than others, but all must be considered in finding the loan arrangement that best suits your needs.

1. Primary Elements

Three essential elements in obtaining a loan and their inherent relationship are illustrated in this little scenario:

You're looking for a $100,000 loan, so you turn to the financial section of your paper and see Dandy Don's Loans advertised. Naturally, you're looking for the lowest interest rate you can find, and Dandy Don advertises a seven percent loan. So you call Dandy and say, "Dandy, I see you've got a seven percent loan."

"Yep, sure do."

"Well, where are you located?"

His office is about sixty miles away, but his rate is good so you get in your car and drive sixty miles to discuss a loan with him. When you arrive you sit down and ask, "Is it really seven percent?" And he says, "Yes, it is."

"Well, what else do I need to know about it?"

"It's a three-year loan and the payments are $3088 per month."

Realizing that this loan would be impossible to handle on a monthly basis, you go home and proceed to look for a loan which would be acceptable. You look in the paper and notice Charlie's Loans. Charlie has a nine percent loan, two percent above Dandy Don's seven percent, but Dandy Don's seven percent loan was worthless because of the terms and you think maybe a nine percent loan will have better terms. You call Charlie and say, "Charlie, what's the term on your loan?"

THE ESSENTIAL ELEMENTS TO CONSIDER IN SEEKING CONTRUCTION AND PERMANENT LOANS

    I. *Primary Considerations*
        1. Rate
        2. Terms
        3. Points
        4. Settlement Costs
    II. *Often Overlooked Items*
        5. Owner/Builder
        6. Disbursements Schedule
    III. *The Permanent Loan*
        7. Assumability
        8. Prepayment Penalty
        9. Loan-to-Value Ratio
    IV. *Miscellaneous Considerations*
        10. Borrow the Money That You Need
        11. Look for the "Friendly Banker"

12. Reputation and Stability of the Lender
13. Turnaround Time (There are two of them)

"It's a thirty-year loan."

You breathe a sigh of relief.

So you go see Charlie. You checked on his rate and it's a good rate. You checked on his term and it's a manageable term. So you sit down, and say, "Now Charlie, is there anything else I need to know about the loan?"

And he says, "Yeah, have we talked about points?"

You say, "No."

Now you need to understand what points are. Points are fees charged the borrower by the lender for the lender to arrange or originate the loan. A point is one percent of the loan amount, or $1,000 for every $100,000 of the loan.

In this case, in order to get you a nine percent, thirty-year loan, there are fifteen points.

You say, "Ouch."

This imaginary scenario illustrates an essential aspect of seeking a loan. You must take into account all the elements involved—the interest rate, the terms, and the points. You may find one lender offering a fantastic rate but his terms of payment or his demand for points might be unacceptable, or vice versa.

a. RATE, TERMS, POINTS

The cost of the loan to you over a fixed period of time will be the result of a mixture of rate, terms, and points. These often vary from lender to lender and you'll have to decide what is most important to you. Some people can't accept the thought of points. The main consideration, however, is going to be the time frame. The longer the term of the loan the lower you want your rate of payment, even at the risk of having to pay higher points. The shorter the term, the lower the points should be. You can afford a higher rate for a shorter period of time. You must weigh these considerations and come up with a combination of rate, terms, and points that is best.

On a new acquisition, points are tax-deductible, but not on a refinance. When you settle on a real estate loan (construction or permanent) you must pay the points at the time of settlement. If you borrow $100,000 and you have two points, you need to write a check for $2,000 at settlement, and the points are then tax-deductible. If you borrow $100,000 and the lender nets out the points (meaning that you only receive $98,000) then you have not paid the points. You've borrowed $98,000, and you're going to pay the points over the life of the loan.

The bottom line on rates, terms, and points is that there are trade-offs. A decision concerning a loan should be based on a variety of factors.

b. SETTLEMENT COSTS

An extremely important consideration when you are building a home is the settlement costs.

There are two categories of settlement costs: major and minor. Both costs should be analyzed carefully, especially from a comparison standpoint.

The major settlement costs are points, recording fees, title insurance, and attorney's fees.

(1) *Points.* Points are the fees you pay the lender for originating (or making) your loan. A point is one percent of your loan amount. The points you pay for your loan can vary from zero to six. They are a significant portion of your loan and settlement costs.

(2) *Recording Fees.* Fees for recordation are the fees you pay the county and/or city to record your loan documents for public record. These fees can vary dramatically from region to region. They vary from a few dollars to two percent of the loan amount. Unfortunately, they are not a negotiable item.

(3) *Title Insurance Fees.* The title insurance fees include the cost of checking the title and the premium for issuing the policies. The premiums are regulated. The title search can be expensive and is highly negotiable.

(4) *Attorney's Fees.* Attorney's fees are professional fees you pay your attorney to handle your settlement. Those fees can be small or large. Attorney's fees can range from less than $100 to several thousand dollars. The fee you pay depends in part on how complicated your settlement is and hard you negotiate. They are highly negotiable.

The minor settlement costs are numerous. Let's look at some of them:

(1) *Application Fee.* The application fee is what the lender charges you to apply for the loan. This fee ranges from zero to a few hundred dollars.

(2) *Credit Report Fee.* The credit report fee is what you pay the lender to obtain your credit report. The fee ranges from $25 to $100. It's not negotiable.

(3) *Inspection Fee.* A fee charged to inspect your home during the construction period. This fee can range from zero up to $1,000.

(4) *Surety Fees.* Some lenders will require a surety bond to insure the completion of your project. A surety fee can range from one half to one percent of your loan amount.

(5) *Notary Fees.* These should be just a few dollars.

There are other minor fees which we feel require a second or closer look. Some of these are just disguised money-makers:

1. Settlement Services
2. Special Services
3. Amortization Schedules
4. Underwriting Fees
5. Tax Services
6. Warehouse Fees
7. Loan Processing Fees
8. Photography Fees
9. Transfer Fees

In addition to these major and minor categories, remember that you will first have a construction loan and then a permanent loan. In most cases you will also have two settlements, although the settlement cost on the permanent loan will be significantly less than your construction loan because some of the costs are duplicative or combined.

In some cases, however, you can find a lender that offers what is called a single-settlement loan. A single settlement is one settlement for both loans. In most cases, this single-settlement loan will cost less than a two-settlement loan. Accordingly, it would be a preferred loan from a settlement standpoint. A word of caution, however. Although a single-settlement loan should cost less than two settlements, it may not. The important comparative consideration is not one or two settlements, but how much?

2. Often Overlooked Items

a. *OWNER-BUILDER*

As a matter of policy many lenders will not lend construction money to individuals who have not had previous homebuilding experience, or who will not agree to hire an experienced builder. It is essential that you let your lender know early if you plan to be your own contractor in the building of your home. If you don't make this known at the outset it's entirely possible that you could spend a lot of time and perhaps money in obtaining an excellent loan, only to find out just prior to settlement that the lender will refuse to make the loan when he learns that you have not hired a licensed contractor. So under no circumstances should you attempt to hide this fact from the lender.

The lender might say, "There is no way we can make a loan to an owner-builder who has had no building experience." Or he might say, "We're not sure we can make a loan to an owner-builder who has had no experience."

If you can demonstrate some knowledge of how you're going to go about building your home, the lender might say: "Okay, we'll take the risk." The more knowledgeable you are about homebuilding the more likely a lender will be to loan you the money.

There are seminars on homebuilding available. And in some cases, lenders will accept certificates from institutions offering these seminars as evidence that you have received some instruction in homebuilding. We hope too that the knowledge gained from reading this book will help you convince a lender that you know how to manage the building of your home.

b. DISBURSEMENT SCHEDULE

The disbursement schedule is the manner in which the construction loan funds are advanced to you as you make progress on the building of your home. Money is the oil that makes your project proceed properly, and the more readily accessible the money the more smoothly your project will move ahead. You will have to purchase material, pay contractors and pay fees as you go along. If you don't have the money to take care of these as you need it your project can be held up. If you are in a strong cash position the disbursement schedule is not important. You can use your own funds to keep the project going and replenish your funds at an appropriate time during the disbursement process. However, if your working capital is thin you should find a lender who will offer you a fluid and flexible disbursement schedule.

There are essentially three types of disbursement schedules. One is a very stingy disbursement schedule, and there are two aspects to it: (1) the lender stays ahead of you, in that the amount he advances is much less than the work you have done; (2) the lender does not inspect the site and disburse funds frequently. They may disburse only three or four times throughout the life of the project. If the lender requires that a great deal of work be completed before advancing money, and imposes a lengthy period of time between when the work is done and when you get the money, you're saddled with a stingy draw schedule.

The standard draw schedule which the borrower will see most frequently is one consisting of between five and seven draws. A five-draw schedule is shown in the illustration. You may find one with six or one with seven. You probably will be able to live with this type of draw schedule. You may also be able to negotiate with the lender and move certain items back and forth between the particular draws in the schedule.

The third type of disbursement program is one which is fluid and flexible. This is the one we prefer, and it can take two forms. In one version the lender will predetermine percentage values for various components of the home (as in the schedule we show here). The lender will say, "We'll give you as many inspections as you like as long as you pay for them. You can call for them at any time, early or late. When we come out we'll inspect the house and then add up everything that's been done and multiply it by these percentages. Then we'll multiply the total percentages by the amount of the loan and we will disburse that money to you."

This is a typical flexible-draw schedule. In terms of a grading, we give it a B +.

The second form of the flexible draw schedule is one on which the lender will pay almost by invoice. You have a well drilled, submit the invoice, and the lender will pay for it. The same thing goes for the septic system, the foundation, the floor system, the roof system, the windows, the cabinetry and so on. This is an optimum type of draw schedule.

A STANDARD CONSTRUCTION DRAW SCHEDULE

The Lender is authorized to disburse the net proceeds of the loan in accordance with the following schedule, provided the work is approved and disbursements authorized by the Lender's inspectors.

$ _____ (20%) First house construction advance: When footings, foundation, first floor joists and subfloor are complete (in the event of a slab foundation, when slab is poured and all plumbing roughed-in.)

$ _____ (20%) Second house construction advance: When house is under subroof with roof sheathing and exterior wall sheathing installed; all interior partitions are roughed-in.

$ _____ (20%) Third house construction advance: When roofing, chimney, siding and windows are in place; plumbing, mechanical and electrical rough-in work is complete; insulation and drywall is installed and basement floor is poured.

$ _____ (20%) Fourth house construction advance: When staircase, interior and exterior trim, casework and countertops are in place.

$ _____ (20%) Fifth house construction advance: When bathroom fixtures, interior and exterior painting, heating and/or air conditioning system and water and sanitary facilities are complete; appliances, finish flooring and carpet, light fixtures in place; driveway and finish grading and landscaping are complete; and, the Lender has been furnished with a complete Release of Liens for all material and labor on-site and off-site.

$ _____ TOTAL

All disbursements are to be made payable to:

The most important thing about these disbursement schedules for you as borrower is to make sure (1) that you receive money for work done, essentially on a dollar-for-dollar basis, (because if you don't, you'll be constantly behind) and (2) that there will be as many inspections and subsequent disbursements as possible and as frequently as possible. Don't hesitate to ask the lender about this and discuss it with him because the disbursement schedule is critical to the success of your project.

There are two disbursement philosophies that guide most lenders. One is: "Stay way ahead of the customer. Get the customer in a situation where he has done more work than you have advanced money for." This is the ultimate protection for the lender. But you should avoid a lender with this disbursement philosophy or you will end up working for him. It will be very, very difficult for you to complete your project. Your lender will be well-secured, but your project will be bogged down in mechanics.

The second philosophy is that since inspections cost money, if the lender limits the number of inspections he can save not only the customer money but himself as well. This is fine for the lender, but it can make it difficult for the owner-builder to get his work done on schedule. You could probably get the lender to increase the number of inspections, however, if you're willing to pay for them. It will be helpful to identify the lender's draw philosophy.

In summary, you need to know how frequently the lender is going to inspect what has been done and how soon after the inspection you're going to get your money. Does your lender inspect on reasonable notice within a matter of days, and will he write you a check within a matter of days? These are relevant questions, and your decision whether or not to accept the loan should hinge on the answers. For some owner-builders, it may well be the most important element of a construction loan.

3. The Permanent Loan

a. ASSUMABILITY

Under most financial conditions, it would be desirable to have a permanent loan that is assumable. This means that you can pass the loan on to anyone who buys your house—if the buyer can qualify for the loan from both an income as well as credit standpoint. When the interest rates are high, an assumable loan is not very desirable, because no one wants to take over an eighteen percent loan after the rates had gone down to around eleven percent. In a low or moderate interest-rate market, when interest rates apparently are rising, if you have a nine percent, ten percent or eleven percent loan that you can pass on it's attractive to a buyer. Assumability is not a factor of major importance, however, so don't turn down a good loan just because it's not assumable.

b. PREPAYMENT PENALTY

A prepayment penalty is a charge for paying off a loan sooner than the note calls for. Historically loans have frequently carried prepayment penalties. But as a result of public resistance and subsequent pressure on the U.S. Congress and state legislatures, prepayment penalties are now a thing of the past. If you see a prepayment penalty there will probably be three important limitations on it: (1) it will never be more than two points, (2) it will lapse after a five-year period, and (3) it can never be more than a prepayment penalty on the first $75,000 of your loan. Given these restrictions the prepayment penalties are not really a problem. Accordingly, do not reject an otherwise attractive loan simply because it has a prepayment penalty. In all probability you will never be confronted with the penalty, and if you are it's a very small price to pay for an otherwise attractive loan.

c. LOAN-TO-VALUE RATIO

The loan-to-value ratio is the amount of money a lender will lend you based on the value of your project. There are two things to consider here: (1) If you are building a $100,000 project, and you have $10,000 in cash, you need a ninety percent loan. Some lenders will not make a ninety percent loan. As a result, they are not viable candidates to be your lender. Obviously you need to find a lender whose loan-to-value ratio will be sufficient to bridge the difference between the amount of money you need and the amount of cash you have. (2) The higher the loan-to-value ratio, the higher the risk is to the lender. The higher the risk to the lender, the higher will be the rate to the customer. For example, if your lender makes an eighty percent loan at a ten percent rate, he will charge you more than a ten percent rate to make a ninety percent loan, and higher still to make a ninety-five percent loan. The greater the risk, the higher the rate.

The question frequently asked of lenders is: If you will make me an eighty percent loan at ten percent, and a ninety percent at ten-and-a-half percent, what will you do for me at fifty percent? Will you give me a six percent rate? The answer, categorically, is no. Once a lender determines an acceptable minimal risk level, usually seventy-five to eighty percent, any ratio less than that does not make him feel any safer about the risk, and at that particular point he's dealing with the cost of funds and profitability.

There are areas in the country where savings rates are very high relative to building activity. And there are areas where the building activity is very high relative to the savings rate. The savings rate in Buffalo, New York, is very high relative to the building rate. The building rate in Tampa,

Florida, is relatively high relative to saving. So what happens? Investment money flows from Buffalo to Tampa.

Whether you want to negotiate a maximum loan-to-value ratio is a personal decision. But keep this in mind. The interest rate on a home mortgage is historically one of the lowest rates you will be charged on any type of loan. Hence, as a general rule we recommend that you borrow the maximum amount possible that you can be comfortable with.

4. Miscellaneous Considerations

a. *BORROW MORE MONEY THAN YOU NEED*

It's usually a good idea to borrow more money than you need in your construction loan. If you don't spend the money you don't have to pay interest on it. Furthermore, when you convert to a permanent loan from your construction loan, you can convert for a lesser amount of money and, accordingly, a lower payment rate. It's always easier to go from a high construction loan to a low permanent amount than the other way around. When you move from a lower construction loan amount to a higher permanent loan amount you have to go through the whole settlement process again. But you don't have to do so when you convert from a higher construction loan amount to a lower permanent amount.

Incidentally, if you run over the figure you have borrowed, don't despair. Most lenders will make you a second mortgage.

Invariably you will find, as you come down the stretch, that you will need more money than you arranged to borrow. Construction money is a lot cheaper than MasterCard or Visa loans. But there is one negative side to borrowing more money than you need. If you estimate that your house is going to cost you $100,000 to build and you make a construction loan of $105,000, you have to pay points on the full amount—though you pay interest only on the amount borrowed.

b. *LOOK FOR THE "FRIENDLY LENDER"*

Try to find out how easy it is to deal with your lender. Before it's over you're going to have to establish a rapport with him. If he's difficult to deal with or the two of you have a personality clash it can be a problem. You want a lender who is a team player. The two of you are a team involved in building and financing your home and you want somebody who will say, "Sure, I'll be glad to come out and see your house on a Saturday afternoon." Or, "I'll be glad to meet you one evening to talk about changing the draw schedule." Your lender should be someone who is willing to go a little extra distance to accommodate your building needs. Fortunately, there are a great many lenders who will do so.

We know of lenders who have made personal deliveries of checks to customers on Saturday and Sunday afternoon after having made a late inspection on a Friday night. We have also known of lenders who, when the borrower has requested money for particular work on a Thursday and the lender cannot inspect until Monday, have actually mailed the borrower a check on Thursday so he could have it on Friday, and then gone out on Monday to inspect the work for which they had already advanced the money. In this case the lenders obviously had confidence in the honesty of the borrower. It's not a common occurrence to be sure, but it's how things can work when you have a friendly, cooperative lender. If you're not comfortable with your potential relationship with the lender you may want to keep looking. This relationship is as important as the rate.

c. REPUTATION AND STABILITY OF THE LENDER

Lenders, unlike most banks, do come and go. You should make sure that your lender is going to be around for a while before you begin doing business with him. Lender is a very broad term. All banks are lenders but not all lenders are banks. Perhaps seven out of ten lenders will be banks. There are full-service banks, savings and loans, and mutual savings banks. Many people will shop for a construction/permanent loan with a mortgage company, however, and mortgage companies do come and go. If you're halfway through your project and your lender goes bankrupt or out of business, you will not be able pay your contractors. This could create a serious and perhaps disastrous problem for you. There have been any number of cases in which builders and individuals were well into a project and suffered egregiously when a lender went out of business. In the recent Ohio and Maryland savings and loan crises numerous builders and owner-builders suffered. A new lender does not like to come in when a project is half done. You'll have to have a new credit check and the project will have to be appraised and inspected all over again.

Once you've secured a permanent loan there is no problem. At that point you only obligation is to pay the money. Thus the stability of a lender on a permanent loan, once it is funded, is not important. You are interested in the stability of a lender during the construction period because that's when you should be working together as a team, and you're relying on him to pay you once you've performed on your part of the contract. So look into the reputation and stability of the lender before you borrow money from him.

d. TURNAROUND TIME

We know of lenders who take from six to twelve weeks to process a loan and get ready for settlement. Some lenders can do it in two to three weeks.

Once you're ready to build it's important that a lender conform to your building schedule. Ask him what his turnaround time is and make sure you get the right answer. All too frequently people underestimate both the time and effort involved in securing a loan. Don't wait until the last minute to apply for your loan.

There are actually two turnaround time periods: loan approval and loan settlement. Accordingly, the relevant question for your lender is, "How long before you approve my loan and how long after approval before I can settle on my loan?"

C. THE CONSTRUCTION LOAN

A construction loan is a loan that enables you take the house from concept to blueprints to occupancy. A construction loan has a very short term, generally from six to twelve months. If for any reason you don't finish the home during that period, the time usually can be extended. The smaller the home and the more favorable the climate the shorter the time period needed for building a home; thus the shorter will be the time you'll need a construction loan for the completion of the home. If you're building a large home in an area where you can expect bad weather, or if you have a particularly difficult job site that will involve extra time on the foundation, you'll want to look for a construction loan that will give you at least nine (and preferably twelve) months in which to build a home. For a number of reasons, it may take you longer to build a home than you plan, and you should plan for that eventuality. You're much better off with a nine-month construction loan and finishing the home in six months, than with a six-month loan and finishing it in nine months.

Terms are therefore an important consideration, and you'll want to maximize the length of the loan. Make it as long as possible. Usually you can obtain longer terms without giving up other items.

The construction loan can have either of two interest rates: fixed or variable. But don't place too much emphasis on either type of rate of the construction loan or the rate itself, because you'll be borrowing this money for only a very short period of time.

The rates of a loan are usually determined by risk. For the lender the risk is greater during the period of a construction loan than for a permanent loan; hence, the rates for a construction loan are usually higher. A house under construction is an architect's concept, a builder's plan, and a piece of property. There are risks associated with taking it from conception to completion. Many lenders do not even make construction loans; those who do constitute a very narrow segment of the market.

Your construction loan could have a fixed rate or a variable rate tied to the prime rate.

Normally a construction loan is a variable rate structured on the formula of prime plus a percentage. The percentage over prime is a negotiable item.

We have seen how a construction loan works. As you make progress on your home the lender inspects the home and advances the money accordingly. At the end of the month you're charged interest only on the amount of money the lender has actually advanced. Obviously, as you make progress on the home, the amount of money that you've borrowed will increase, and the interest payments at the end of the month will be larger. If you obtain a loan for $100,000 and you use $10,000 of that loan in the first month, then you'll be paying interest only on the $10,000. If your interest rate is twelve percent, you have to pay back one percent of the $10,000, or $100. Obviously the interest rate on the construction loan is not something to be overly concerned about, as it will comprise a very small part of the total cost of the project. But if paying this interest is going to be a problem, the interest charges for the construction period should be added into the budget of the loan so that you can draw upon that total budget in order to service the interest during the construction period.

A construction loan is paid back either when you secure a permanent loan or when you convert from construction loan to a permanent loan (which you would do in a single-settlement loan). If you have a single-settlement construction loan, the construction loan rolls over into a permanent mortgage and the construction loan is paid off over time via the permanent mortgage.

D. THE PERMANENT LOAN

The permanent loan is the loan that will enable you to stay in the house for a prolonged period of time and retire the debt by making monthly payments on the mortgage. In recent years there has been a wide variety of options for the permanent loan. The Federal Reserve Board has published a pamphlet explaining approximately ten of the permanent mortgage options that were available.

Today, there are essentially two options concerning the permanent loan: the fixed-rate, fixed-term, and the adjustable-rate mortgage (ARM).

1. The Fixed Rate

Historically, the most common and most preferred loan is the fixed-rate loan. You borrow at the then-prevalent interest rate, and you pay at that rate until the loan has been liquidated. The fixed-rate loan usually comes with two choices: the classic thirty-year fixed-rate, fixed-term loan, and the relatively newer fifteen-year fixed-rate, fixed-term loan. The distinct advantage of the thirty-year loan is that it gives you stability and security for thirty years. Because the principal of the loan is spread over a long period of time, the monthly payment on the loan will be less than the monthly payment on a fifteen-year loan. This difference is important because you qualify for a loan based on the monthly payment of the loan as a percentage of your income. Accordingly, the lower the monthly payment, the larger the loan for which you can qualify.

Interest paid on your home is a tax-deductible item; hence, one advantage to the thirty-year loan and its usually higher interest rate is that it will give you a larger annual interest deduction on your IRS form.

The fifteen-year loan is obviously a lesser risk to the lender. One advantage of the loan is that, because the risk is less, the interest rate will be lower. You should expect a rate differential of one-quarter to one-half percent between the fifteen-year loan and the thirty-year loan.

The second advantage is that, because of the shorter term of the fifteen-year loan, you will build up equity in your home sooner. In fact, because of your larger monthly payments, you will have a faster equity build-up even if your interest rate remains the same.

However, the disadvantage of the fifteen-year loan is the higher monthly payment schedule which makes it more difficult to qualify for than the thirty-year loan.

There are other terms for a fixed-rate loan. Sometimes the term is ten, twenty, and twenty-five years, and in rare instances, forty years. But probably ninety percent of all the fixed-rate, fixed-term loans are thirty years. Choose the term with which you are most comfortable.

2. Adjustable-Rate Mortgages (ARMs)

Although the fixed-rate, fixed-term loan is the most popular loan because it is a stable, known quantity, it is not necessarily better. For many borrowers the new adjustable-rate mortgages (ARMS) may be the most advantageous way to structure a permanent loan.

The ARM is relatively new on the financial market and is a direct product of the high inflation rates we experienced for a short time in the early 1980s and the banking deregulations that occurred during this same period.

When interest rates took off in the 1970s, many lenders found they were holding long-term (up to thirty-year) loans at around nine or ten percent or less at a time when current interest rates were as high as eighteen to twenty percent. This created problems for the lenders. Even after the rates returned to what we might call normal figures, bankers began to think about ways in which they might protect themselves in the future if inflation drove the interest rates up again. They decided that an adjustable rate geared to some barometer of the interest rate, such as the Federal Home Loan Bank Board Index or the price of government treasury bonds, would protect them. Once they came up with this concept they were able to lower their rates dramatically, because by reducing the term of the loan the lenders reduce their risk. With the lower risk came the lower rate.

Most people want a long-range, permanent-rate loan for the same reasons the banks don't like them. It freezes the rate for the next fifteen to thirty years, no matter how high the interest rate soars. The fact is, however, that most people never pay out a thirty-year mortgage. Few people will make the 360th payment; most people probably will not even make the 120th payment. The chances are all that they will either sell their home or refinance it before that time. Statistics show that if you live in certain areas, in all probability you will not even be making the 90th payment, which is in seven and a half years. So if you are not going to be making the 90th payment, why be concerned what the rate is beyond the 90th month?

In determining whether an ARM is a good mortgage for you, there are six factors you must consider:

a. The initial rate
b. The adjustment interval
c. The maximum rate increase per interval adjustment
d. The maximum lifetime rate increase, called a cap
e. The index to which the rate is keyed
f. The rate above the index rate

a. *THE INITIAL RATE*

In order to make their loan more attractive most lenders will offer an introductory rate which is frequently well below the going rate for a short time, usually your first interval period. However, the initial regular rate will also probably be below the going rate for fixed-rate mortgages, and this initial rate is important. It will usually represent the minimum low you will experience during the term of your loan. When added to your lifetime CAP, this will give you the maximum rate you can experience during the term of the loan. And if you add your initial rate and the maximum rate and divide by two, theoretically that should give you the average rate and the average cost of the funds during the lifetime of your loan. This rate should approximate the going rate for the fixed thirty-year loan.

If for any reason this rate is less than the thirty-year fixed-rate loan the ARM is clearly an attractive loan. If for any reason it is significantly higher, then you should take the fixed-rate loan—assuming that you can qualify for it. For example: assume that you have a 7½% ARM with a 5% lifetime cap. That would give you an average of 10%. If the available fixed rate is less than 10%, the fixed rate would be a better bargain. If the fixed rate is greater than 10%, the ARM would be better.

b. THE ADJUSTMENT INTERVAL

The adjustment interval is how often your rate will be adjusted. The interval will generally be in the range of every one to five years, or even as frequently as every three to six months. As you go from a thirty-year rate to a rate for one year (or less) the risk on the part of the lender is obviously progressively much less. So the more frequent the interval, the lower the risk, which means the lower the rate.

Below are some typical lenders' interest totals which compare the rates for thirty years down to one year.

YEARS	30	15	5	3	2	1
RATE (PERCENT)	10	9¾	8¾	8½	8¼	7½

This table shows that there is a direct correlation between the terms of the loan, the risk, and the rate. The frequency interval is important. The greater the time frame between the rate adjustments, the higher the initial rate will be. If you select a one-year ARM, the initial rate should be lower than if you select a five-year ARM. The frequency interval will also have an effect on your initial starting rate. In short, you pay for stability.

The ARM rate schedule is available in the financial sections of most newspapers, and you can also ask your lender for a breakdown of his loan terms and rates. If they do not show a logical ratio of term to rate, then something is wrong—although any discrepancy may be in your favor.

c. MAXIMUM RATE INCREASE PER INTERVAL

Most variable-rate mortgages for fixed amounts limit the extent to which the interest rate can rise in any given interval. This limit is usually governed by fractions, such as ⅖. In the case of a ⅖ fraction the numerator (2) is the maximum rate increase of your loan in any given interval and the denominator (5) is the maximum rate increase of the loan over the life of the loan. In most ARMs the rate can go up or down throughout the life of the loan, although the rate for most ARMs will never drop below the initial rate charged at the time you borrowed the money. It is important to remember that ARMs do come down! Once they have risen above the base rate they can drop down again.

d. THE MAXIMUM LIFETIME RATE INCREASE (CAP)

In the fraction ⅖, the five means that the loan interest rate can only increase a maximum of five points above the starting during the life of the loan. So if you begin at a 7½% starting rate, and your adjustment interval is one year, the worst-case scenario at the end of the first year (using the ⅖ ratio) would be 9½%. At the end of the second year, it would be 11½%; at the end of the third year the maximum it could be would be 12½% which is the maximum cap of your loan. But what is the equal distance between 7½% and 12½%? Ten percent. We are still shopping now, and if the fixed rate is below ten percent, then the fixed rate would probably be better, because we assume that your loan is going to fluctuate on the average between 7½% and 12½% which is ten percent. So in order to beat the fixed rate, you have to average less than ten percent. But if you stay below ten percent early in the life of the loan as opposed to later you are ahead of the game. Why? Because you will be borrowing more money early than you are later. The statistical likelihood of your being above the fixed rate for thirty-six months is very low. So this would not be a bad loan.

You may also see an ARM loan which is based on the fraction ⅙. And here is what might happen with the fraction: You start with a 7½% rate and taking a worst-case scenario, at the end of one year you are at 8½%; at the end of two years you are at 9½%; and at the end of three years and 10½%, up to a maximum of 13½%. You may have a choice between 7½% and a ⅖ cap and 8% (as a starting rate) and a ⅙ cap. Which one would you choose?

You need to look at all the options and to make an effective decision. Sometimes you will see fractions that govern your loan based on 1% or ½% or ¾% as the numerator and infinity as the denominator, meaning that there is no lifetime cap on the loan. If so, don't despair, because the advantages of either the starting rate or the smallness of the frequency adjustment may be so attractive as to offset the fact that there is no lifetime cap. For example, you may get a loan in today's market at 6% starting rate, and if it were limited to a ¾% increase per year with no lifetime cap, it would still take eight to ten years before your interest rate would be one that would be unacceptable.

e. THE INDEX TO WHICH THE RATE IS KEYED

Lenders use definite guides and regulations to determine the up and down movement of your interest rate. The most common index used to determine the rate is the treasury bill—T-bill. If you select a one-year ARM, then your loan will likely be keyed to a one-year treasury bill. If you select a three-year ARM, your loan will be keyed to a three-year treasury bill. At the time of your adjustable interval the current T-bill rate will determine the interest rate on your loan until the next frequency interval. Another common key used is the Federal Home Loan Bank Board Index, which is a composite of all mortgages in the secondary market in the United States.

There are, however, a few loans that may be keyed to neither of these standard indexes. They may be keyed to the lender's own index, which will not offer the same type of stability and certainty that more common loans do. This may not be a bad loan, but it is one which deserves careful scrutiny.

f. *THE RATE ABOVE THE INDEX RATE*

It's important to understand that your rate on the ARM is not the rate on the T-bill, or the Federal Home Loan Bank Board Index, or any other index that might be used. You always pay a spread above the index. The spread above the index is generally somewhere between two or three points. That spread, which determines your final rate, may be a negotiable item. You should shop for the spread above the index as diligently as you do with the other elements of your loan.

g. *IS AN ARM GOOD FOR YOU?*

It's not easy to determine whether an adjustable-rate mortgage is right for you. Even if you decide you want one, banks are offering so many ARM options that it's very difficult to determine which is best. But if you think you will be among those who will refinance or change houses in the next eight years (and you would be surprised how many people are ready to sell their dream house when they realize it has doubled in value in five years), then you might want to consider an ARM. Another distinct advantage of the ARM is that, with its lower initial rate, you can qualify for a larger loan than you would qualify for with a fixed-rate loan. In some cases the ARM represents the only option for obtaining the loan you need to build the house you want.

E. OTHER PERMANENT LOAN OPTIONS

In addition to the conventional thirty- and fifteen-year permanent fixed-rate, fixed-term loans, there are a number of other options which lenders and real estate people sometimes call exotic loans. Here are a few of them.

1. The Biweekly Loan

One advantage of this loan is convenience. The payment date corresponds to your payday. This loan is also frequently promoted because it pays off more quickly than the typical thirty-year loan. There's nothing magical about this. The reason it pays off more quickly is simply that during a one-year period, you will make thirteen payments rather than 12. So a more accurate description would be to say that it falls somewhere between a thirty-year loan and a fifteen-year loan. Even if the loan is perfect for you, it is probably not a loan that most of your prospective buyers for your house would want to take over. We recommend that it be avoided, unless the convenience of the coincidence of your paycheck and your mortgage payment is important to you.

2. The Graduated-Payment Mortgage (GPM)

The graduated-payment mortgage is different from an adjusted-rate mortgage. In the case of the ARM, your rate adjusts. In the case of a GPM, not only does your rate adjust in accordance with market conditions, but your payment will change over a certain time period. This gradual increase in your payment will enable you to catch up with any deficit you may have in the amortization of your loan. What usually happens with a GPM is that you start out with an initially low payment, but over the first five years of the loan the payment (not the interest rate) will increase by some pre-determined amount per year, usually from five to seven and one-half per-cent. Because of the complexity of this program, the rate is higher on balance than a fixed-rate, fixed-term loan, and frequently the ultimate rate is higher than on an ARM. The advantage to the loan is that it enables you to qualify initially because the payment rate is lower initially. And it is an-ticipated that your income will go up as your interest rate and monthly payments increase.

3. Shared Equity Mortgages

Even if you are building your own home, you may want to consider bring-ing in an investment partner who will perhaps put up the down payment for the project or produce some equity and share in a percentage of the ownership (frequently one-half). As a rule, you make the monthly pay-ments on what could be almost any kind of a fixed-rate or an ARM loan. This is a relatively new type of financing. Somebody puts the money up and you make the payments on it. The person who puts in the money is comfortable with his investment because you are making the payments and have a vested interest in the home. So the investor knows you are going to make the payments and take care of the property. Usually the shared eq-uity mortgage is reviewed at the end of some time period (generally five years). At that point the home is appraised and an option is available to either party to buy the other one out.

4. The Balloon or Bullet Loan

The balloon or bullet loan is a loan which is amortized over a longer period of time, but matures over a shorter period of time, with one large payment due at the end of the loan period. For example, you can have a thirty-year loan that balloons in five years; a ten-year loan that balloons in three years; or a fifteen-year loan that balloons in ten years. The reason for the longer amortization is to keep the payments down. The reason for the short-term maturity is to minimize the risk to the lender. Generally balloon or bullet mortgages have lower rates because the actual term the money is loaned out is shorter.

There are some attractive balloon or bullet loans available. One particularly attractive one is amortized for thirty years, but does not balloon for ten years. This is particularly attractive because people are rarely in their home for a ten-year period. For those who feel that they will not be in their home for longer than ten years this is a particularly attractive loan. The payments will be on a thirty-year basis, but since the risk to the lender is only for ten years the rate should be lower.

One concern for the borrower is: what happens at the end of ten years? Unlike the ARM, there is no provision in this loan for refinancing at the end of the ten-year period. You will usually find a rate that will fall somewhere between a five-year ARM and a fifteen-year fixed-rate loan.

F. SPECIAL LOANS

The loans we have discussed so far are those that enable you to build your house and enable you to occupy it over a long period of time. They are usually called first-trust loans because they have first consideration over any other loan that you might negotiate and for which you use your house as collateral. But there are some special loans with which you should be familiar.

1. The Second-Trust Loan

A second-trust, or second-mortgage, loan is also secured by your real estate, but is in second position to the first trust. When you are building a home, you may find that near the end of the project you want to add a garage or a swimming pool. Or you may find that you are short of funds. You have two choices: you can either refinance your first trust, or you can approach your lender or another lender about a second trust. Second-trust loans are generally for a shorter term and are at a higher rate, because they are a higher risk. If you fail to make payments and the house is sold, the first trust balance must be paid off before any proceeds go to pay off the second-trust loan.

2. The Bridge Loan

If you own an existing home but want to build a new home without selling and moving out of your old home before the new one is ready, you will need cash to finance the building of the new home. If you have ample cash in the bank, there is no problem. But if you are like most people, your main source of funds will be the equity you have in your existing home. If that equity is substantial, you can obtain what is called a bridge loan.

The bridge loan is often a second trust, although it does not have to be. It could be a third trust. The importance of a bridge loan is that it enables you to stay in your existing home while you are building your new home. When your new home is completed, you sell your existing home, retiring all of the existing trusts, including the bridge loan. The advantage of a bridge loan is that it means you do not have to refinance or sell your existing home and move into a rental home while your new home is under construction. A bridge loan is relatively easy to obtain.

3. The Blanket Loan

A variation on a bridge loan is the blanket loan. A blanket loan is a single loan for a single amount, secured by two pieces of property. The lender makes you a loan secured by the new project and the equity in your existing home. When you complete the new project and sell the existing home, some predetermined amounts from the sale of your existing home will be applied to reduce the principal amount of the loan. The lender ends up with a first trust on the new project and a no-trust position on the old project. You accomplish the same thing with a blanket loan that you do with a bridge loan.

4. The Wrap Loan

If you have a particularly attractive first trust on your existing home and a great deal of equity in it, another option is to take a large second trust, which is called a wrap loan. Generally speaking, the amount of the second trust is higher than that of the first trust. So the lender "wraps" your house in a security instrument, which is a second trust—but the rate on the second trust is a little bit higher than the rate you would get if you were to refinance it. It enables you to retain the interest rate that prevails on your existing first trust. A wrap loan is ideal for someone with an extremely low interest rate on the first trust.

G. LOAN FOR A SECOND HOME

Quite possibly, the custom home you are planning to build is a second home—for recreation or possibly for future retirement. The question is: how difficult is it to obtain a loan for a second home?

The answer is "not very," assuming that you can qualify financially. But there are different conditions under which you might obtain the loan for a second home. First, the loan-to-value ratio on a second home loan will be lower than on a first home. Typically, a lender will make an eighty percent loan on a first home but will only make a seventy-five percent loan on a second home. Why? Because a second home loan is riskier than a first home. The interest rate for the second house loan will also be higher than on a first home. Why? Again, because from the lender's standpoint it is a higher risk. If a borrower has two homes and is in financial trouble he will be more likely to default on his second home than on his primary residence.

There will also generally be a shorter term on a second home loan. Typically, if a lender is writing a thirty-year fixed-rate mortgage for 80% at a 10% rate, the second home will carry a twenty- to twenty-five-year, 10½% loan for seventy-five percent of the value. One new development is that the government, through its secondary markets, will now purchase loans on second homes, which has made it much easier to obtain second home loans.

In summary, refinancing for a second home is readily available but there are three factors that you will need to consider. Generally speaking, the rate will be slightly higher, the term will be slightly shorter, and the loan-to-value ratio will be slightly less (requiring a higher down payment).

H. BASIC LENDING SOURCES

1. The Government

The federal government actually makes very few direct loans. Through its guarantee program, however, it makes the largest amount of mortgage funds available to the market. This is called the secondary market.

The secondary market is created by the government and serviced by quasi-governmental institutions which purchase mortgages generated by the lenders with whom you, the customer, will be dealing. The lender could be a savings and loan, a commercial bank, or a mortgage company. When one of these institutions originates a loan that meets government specifications, it can then sell that loan to the secondary market.

There are several types of government funding under the government program.

a. GOVERNMENT NATIONAL MORTGAGE ASSOCIATION (GINNIE MAE)

Ginnie Mae makes loans available through two different government agencies: the Federal Home Owners Administration (FHA) and the Veterans' Administration (VA). The main advantages of FHA and VA loans are that they require lower down payments and their qualifying requirements are less demanding. However, FHA and VA loans are not practical for the building and long-term financing of a home for three reasons:

(1) There are virtually no construction loans available through FHA or VA;
(2) The red tape and paper work involved is staggering;
(3) The loan amounts available through FHA and VA are generally insufficient for someone building a custom home.

b. FEDERAL NATIONAL MORTGAGE ASSOCIATION (FANNIE MAE) AND FEDERAL HOME MORTGAGE CORPORATION (FREDDIE MAC)

These two quasi-government institutions purchase loans from banks, mortgage brokers, pension funds, and others. They do not originate loans; you can't go into a Fannie Mae or Freddie Mac office and ask for money. What usually happens is that a primary lender such as a saving and loan, a bank, or a mortgage company will originate the loan, take the points, and then sell the loan at a discount to Fannie Mae or Freddie Mac. There is a limit of $187,600 (as of May 1989) on the size of the loan the secondary lenders can buy. Loans in excess of this amount can't be purchased by these quasi-government agencies.

There are also other technical requirements for Fannie Mae and Freddie Mac loans, such as a 28/36 loan-to-income ratio, and a limit of an eighty percent loan-to-value ratio. If your loan does fulfill both the amount and underwriting requirements you have a non-conforming loan and your lender either has to portfolio the loan or sell it to a non-government institutional investor.

2. Non-Government Sources

a. CONFORMING LOANS

Loans that meet government underwriting standards in both loan amounts and in underwriting ratios (to be discussed later) are called conforming loans. They are the cheapest loans because they can be sold easily in the secondary market.

b. NON-CONFORMING LOANS

Loans which are in excess of $187,600 and loans in which the applicant's ratio goes higher than 28/36 percent are non-conforming loans, and they cannot be sold in the government secondary market. There are, however, other sources for funding a non-conforming loan including pension funds, insurance companies, and large financial institutions. These sources buy the bulk of loans that are in excess of $187,600.

c. PORTFOLIO LOANS

Loans which are not sold in either the conventional secondary market or the nonconforming secondary market are called portfolio loans. A portfolio loan could be an important source of funds for building your home. They are so-called because they are booked in the original lender's portfolio of loans. There could be at least three reasons why a lender might not be able to sell a loan to a secondary market: (1) the size of the loan, (2) a certain uniqueness in the loan or the property, (3) a significantly higher debt ratio of the borrower than the ratios under which most lenders will purchase loans. For example, we know of a plastic surgeon who was able to obtain a large loan with loan ratios of 45/63. The 45/63 ratios are much higher than most lenders will accept because it couldn't be sold in the secondary market. However, the particular lender who made this loan reasoned that, with an income of $30,000 a month, even after the plastic surgeon had made mortgage payments of forty-five percent and other debts or payments totaling sixty-three percent (including mortgage) he would still have plenty of money left over for groceries and the movies.

Of course, if you cannot obtain a loan that can be sold in the secondary market, in the nonconforming secondary market, or the portfolio market, there are still other sources, both public and private, for securing a loan.

I. QUALIFYING FOR A LOAN

There are essentially three elements in qualifying for a loan:

1. Credit Rating

You should have good character and creditworthiness. These attributes are important, though not as much so as you might think. We read in the *Wall Street Journal* not long ago about a couple who owned a home on the east side of Houston. With the rapidly declining real estate values resulting from the Houston recession, the day came when they owed the bank more on the home than it was worth on the open market. So the couple just left their home and the bank foreclosed. The couple then went across town and bought another home, qualifying for and receiving a loan from another Houston bank.

How was this possible? The couple had already defaulted on one mortgage loan in the same city. But at that time there were so many defaults and foreclosures in the Houston area that there was hardly time to report bad credit.

There is so much lending and defaulting and bad debt today that large institutions that do a lot of financing have a hard time keeping up with it. So a bad credit record with a department store or credit card company may not necessarily disqualify you from a mortgage loan. But it won't help, either.

2. Equity and Income

The most important requirements you need to qualify for a loan are income and equity in the new home project. The income provision is obvious, and probably the most important, because it tells the lender that you have the ability to make regular payments on a large loan.

Let us say that you plan to build your home on a piece of farmland worth one million dollars, and you own the land outright. So you apply for a one-hundred-thousand-dollar loan to build a house. You have one-million-dollars equity in the land on which you plan to build that house. The lender asks you: "How much income do you have?" And you say, "I don't have a job."

You probably will not get a loan from most lenders. The reason is that the secondary market that buys up most of the loans from the principal lenders will insist that you have an income to service your loan no matter how much equity you have in the home. Incidentally, this one-hundred-thousand-dollar loan, which represents only nine percent of the value of the project, could be made by a lender as a portfolio loan. It is a prime example of how a portfolio loan could work. But remember, when he makes the loan, the lender wants the interest income. He does not want your property or the bad publicity associated with your property going into default.

a. EQUITY

How much equity do you need in a project? Depending upon the amount of the loan and your income in round figures, you will need at least five percent. If you have five percent equity in a project, the loan-to-value ratio is ninety-five. Many lenders will make only an eighty percent construction/ permanent loan. However, there are lenders who will make a ninety percent loan and some lenders who will loan ninety-five percent of the project value.

Suppose you want to build a $75,000 house on a $25,000 lot, for a total project cost of $100,000. You have a good job, so you can qualify for a construction/permanent loan of $95,000. You have a car that is paid for. You don't have a bank account, but you do have some cash. The lot will cost $25,000, but you do not own it. So you sign a contract to purchase the land itself for $25,000 on the condition that you can also get a construction/permanent loan. How much cash do you have to have to qualify for a construction/permanent loan that will enable you to build your house?

The answer is $5,000. You do not have to own your land in order to obtain a construction/permanent loan. All you have to have is enough cash to equal five percent of the total value of the project—or ten percent if the lender insists on a ninety percent loan-to-value ratio.

b. INCOME

How much income do you need to qualify for a loan? Obviously that depends on the amount of the loan. Most lenders will figure your qualifications based on standard underwriting ratios, and the most common ratio is 28/36. The 28/36 ratio means is that up to twenty-eight percent of your gross income (pre-tax) can be used to pay for a mortgage, including taxes and insurance. A total of thirty-six percent of your gross, pre-tax income can be used to service the mortgage, taxes, and insurance and all other debts. If you have an income of $3,000 a month, and you want a $100,000 loan, here is what happens. The lender will calculate twenty-eight percent of $3,000, which is about $840. The principal and interest (PI) payment on a $100,000 loan at eight percent would be about $740. But there are also taxes and insurance (TI). The insurance isn't a large item. Assume that taxes and insurance will come to around $100 a month.

Now we have $840 PITI—principal, plus interest, plus taxes and insurance premiums. From a numerator (28) standpoint, you just barely qualify. But do you qualify from a denominator (36) standpoint? Over and above your house and property payments, you have a car payment of $150 and a MasterCard payment of $90. Do you qualify? Answer: yes. Because if you add those two together, the total is $240, and $240 is eight percent of $3,000. So your ratio on this particular loan would be 28/36. What if you also have a Visa card and owe $500 on that? This puts you over the thirty-six percent figure. Do you qualify in today's market?

That depends. Many lenders will exclude any indebtedness which matures within nine or six months. So if your Visa-card loan will mature within nine months it will probably be excluded from the analysis.

One thing to remember is that if your lender intends to sell your loan in the secondary market, as most lenders do, it must conform precisely to the 28/36 ratio. It is a good idea to do a careful analysis yourself. If you are near that ratio in your PITI payments and auto and charge-card loans, you should try to pay up any loans that require large monthly payments for a relatively small amount of money loaned. If you have a $2500 MasterCard debt on which you are paying $80 a month interest, that loan is likely to run on forever, because your monthly interest payment is high and you are paying off very little of the principal each month. You want somehow to try to pay off that loan, or at least to reduce it to the point where it might mature in six to nine months with modest monthly payments.

A twenty-eight percent (property and house payment)–thirty-six percent (total payments of any sort) is the important ratio. With a ratio of 28/36 or less you will have no trouble obtaining a loan that can be sold to the secondary market.

Let's look at some other typical family situations. A family has a good income and assets, with a ratio of 24/40—the mortgage payment on the loan for which they are applying will come to only twenty-four percent of their income and the mortgage combined with all of their other debts will be forty percent. Will this family qualify for a loan?

It will not qualify for a loan in the government secondary market, and the ratio is stretching it for a portfolio loan. But if the income is significant, as in the case of the plastic surgeon, a 24/40 ratio (although it would preclude the loan from being funded by the government secondary market) could still qualify as a portfolio loan. A 24/44 ratio, however, would begin to strain the limits of the system.

On the other hand, let us consider a family which has a 32/32 ratio. The family's mortgage payment is four percent higher than what the normal underwriting standards would permit. The fact that this family doesn't have any other debt, however, is an assurance that it can probably manage its financial affairs properly. A lender would probably make the loan. But the 32/32 ratio would not qualify in the secondary market, so it would have to be a portfolio loan.

These qualifying ratios are important. The 28/36 ratio is the ratio used by the secondary market. In fact, the 28/36 ratio is rather standard throughout the industry. But there are exceptions. There are many lenders (portfolio lenders) that have ratio requirements of 32/40 and some as high as 34/44. Generally speaking, the higher your income, the larger these ratios can be.

J. WHERE TO OBTAIN A LOAN

In most areas there are several institutions or organizations from which an individual can obtain a loan to build or buy a home.

1. Savings and Loans Institutions

These lenders are the largest source of construction/permanent loans.

2. Savings Banks and Mutual Savings Banks

They are only slightly different from savings and loan institutions.

3. Commercial Banks

Historically, savings and loans made real estate loans and commercial banks have made commercial and personal loans. Now some commercial banks play a major part in residential real estate financing. Most real estate loans are still made by savings and loans associations.

4. Mortgage Brokers

A mortgage broker is someone who will originate a loan for you and place it with a lender, an insurance fund, a savings and loan association, or other institution. Using a mortgage broker does not necessarily mean that you will pay more, because they function essentially like travel agencies; the lenders usually pay their fees rather than the applicants. Mortgage brokers can sometimes find a very good loan and are usually very aggressive in their search because organizing loans is all they do.

5. Mortgage Bankers

A mortgage banker is a lender who uses his own funds to generate loans. He can then either sell the loans or put them in his own portfolio.

We believe that mortgage brokers and bankers are good sources of construction/permanent loans because that is their principal activity, whereas savings and loans associations and commercial banks provide many other functions. Also, mortgage bankers and brokers can "shop" your loan at many lenders—finding not only the best loan but finding a lender who wants you. A word of caution, however. They are not always the best source for construction/permanent loans, and you should shop around before you sign with a mortgage broker or banker.

6. Credit Unions

These are often a good source of money, but they do not always provide construction loans. Rates from credit unions are very competitive because credit unions operate for the benefit of their members rather than stockholders.

7. Pension Funds and Insurance Companies

These are good sources of money, but they usually must be approached through an established mortgage broker.

8. Veteran's Administration (VA) and Federal Homeowners Administration (FHA)

Although these institutions receive a great deal of publicity, they are not very good sources of money. The FHA maximum loan in most areas is $90,000; the VA maximum loan is $140,000. Both will make construction loans but it will take a large station wagon to carry all the forms that you will need to fill out. Most lenders will not make construction loans through FHA or VA, and the law does not require them to do it. Both the VA and the FHA insist on two settlements for construction/permanent loans.

K. HOW TO PRESENT YOUR BEST CASE TO A LENDER

It is important that you be well prepared when you apply for a loan. The following are things you should remember. Some of them may sound silly, but we can assure you they will be noticed by your lender.

- Know your project. Never say, "Well, I don't know" when you're asked a question. You should know your project and be prepared to answer any questions.
- Supply a copy of your deed. If you don't have a deed, bring a copy of the purchase agreement. The deed or purchase agreement identifies the property, and the purchase agreement establishes the price. One of these documents indicates that you either own the property (even if you still owe money on it) or that you are buying it. These documents will be necessary at some point, so you might as well have them now.
- Supply a plat and/or site plan.
- Provide both a narrative and a map to reach the property for inspection purposes.
- Provide a list of all of the contractors on your project, and as much detailed information about each of them as is possible.
- Supply the lender with a budget spreadsheet showing the cost of the project from beginning to end. Include miscellaneous, contingency items, and settlement costs. The more detailed and accurate your budget, the more likely you are to get the loan.
- Bring the most detailed plans you can obtain. An artist's rendering of your home is better than the typical chart of floor elevations.
- Provide a detailed list of specifications. We recommend the FHA 2005 form.
- Bring a check. It will cost somewhere between $200 and $300 to apply for a loan, so when you apply, bring a check. Show that you mean business.

- Bring a credit application. They are easy to obtain. Some loan officers like to fill out the credit application while you are there.
- Bring confidence—in yourself and in your project.

L. HOW TO GET THE BEST RATE

To get the best rate, shop around! Call a number of lenders and ask for rates, points, terms, and so on. Then compare them. When requesting such information, be sure to state the purpose of the loan, and whether you will want a construction/permanent loan or only a permanent loan. When you have chosen what you consider to be the most attractive lender, negotiate! Lenders always sound very specific about their rates and policies as if they were permanently fixed figures, but there are a lot of lenders and all loans are ultimately negotiable.

M. SETTLEMENT COSTS

Most first-time buyers are shocked at how much it costs to settle on a piece of property and surprised at the number of fees, taxes, and payments required. A careful scrutiny of your settlement costs, however, can sometimes save you money. Don't hesitate to question and compare every item. There are several things to look for at settlement time. The following are the most important.

1. Points

The number of points usually depends on the market, but often they are negotiable depending on the size of the lender and the amount of the loan. (Remember, the lender does not get his money for free.) Points pay him for making you the loan and they are a necessary part of obtaining a loan. Lenders generally trade dollars in relation to the interest rate. If they are charging you ten percent for a loan, it is probably costing them ten percent to get the money to lend to you. The points will pay their light bill, insurance, employees, and their profit. Do not begrudge a lender the opportunity to make a reasonable profit.

At the same time, remember there are trade-offs associated with points and rates. If you plan to be in your home for a long time, it is better to pay more points and get the lower rate. If you are going to be in the home for a short period of time (or if you are going to need the loan only for a short period of time) you should minimize the points and accept a rate that may be slightly higher than the current market figure.

2. Recording Fees

These are fees charged by the county or city to record your deed and loan documents. They are not negotiable.

3. Inspection Fees

Most lenders will charge you between $125 and $300 to make the inspections during your building process. If they require $400 or $600 you are being charged more than the industry standards. They are probably including some hidden costs and should be questioned about it. Don't proceed from a combative position; instead, ask, "Is there some way I could save some money in this particular area?" "Could I combine some of my inspections?" "Could you explain to me why it's $600? I spoke with another lender and he charges $200.

4. Legal Fees

Legal fees are very negotiable. After the case of *Goldfarb v. The Virginia Bar Association* was decided in 1969, legal fees for settlement have become competitive, so you should shop and compare. Typically, there is a wide disparity in settlement fees charged by lawyers. Some lawyers still charge the historical one percent of the loan to handle a settlement; but we believe that a charge of one percent of the value of the loan is probably excessive. Other legal fees may be permitted, but $300 to $500 is the most you should have to pay for a lawyer's services. Fees of less than $200 are distinctly possible.

5. Surety Fee

Sometimes, regardless of your financial situation or your or your builder's reputation, the lender may require a surety bond. A surety bond insures the lender that the home will be completed, or that the person who writes the surety bond will repurchase the loan from the builder. With a surety bond the lender essentially has no risk. If you're a financially strong builder a lender will probably make you a loan without a surety bond. If not, the lender may require a surety bond. You generally pay about one percent of the loan for this.

The amount of the surety bond should be for the amount of the loan—not the total price. If you're building a $150,000 house and getting a $100,000 loan, you need a surety bond for $100,000, for which you would pay a fee of $1,000.

You can buy a bond from a bonding company. Some lenders will sell you a surety bond through their service corporations. A major supplier or major contractor on your job is a third source. If you're building a $100,000 home and a major lumber yard is supplying $30,000 or $40,000 worth of material, the lumber yard may be willing to provide the surety bond. You might approach the proprietor and say, "I'm building a house and planning to buy all of my material from you. Will you serve as the surety?" In most cases the building material company will agree to provide the bond in order to get your business.

Remember that surety bonds are also negotiable. If you're borrowing $100,000 and the lender asks for a surety bond, the best response should be, "Well, how about a surety bond for $50,000?" It is unlikely that you could lose the whole amount. Some lenders will agree to this. What is important to keep in mind is that the surety bond protects the lender. If you make mistakes building your home there's not much you can do about it. You're not only out of luck, but you still owe the lender and will also be obligated to pay the surety company.

Go over your settlement sheet carefully, look at your fees, and question any that don't seem right. We know of one case in which a bank asked for a surety fee of $2,625. The borrower called his lender and said he wasn't complaining, that he was sure it was a normal expense, but, "I'm a little short and I don't have $2,625." When the lender went back and recalculated, he found he had made a mistake in the placement of the decimal point and the fee was actually $262.50!

N. THE CREDIT APPLICATION

1. How to Fill Out a Credit Application

Here are a few pointers for filling out your credit application. Proper completion of the application assures the lender that you know what you're doing and understand their needs. The forms may ask for gross, annual, or monthly income. Give the correct figure—do not overstate your income. They usually can catch misinformation, which will not only be embarrassing but may jeopardize your loan.

A frequent question is, "I will retire in six months. I make $58,000 now. In six months, I'll make $21,000. What do I put on the credit application?" You can give the explanation if you wish, but the bottom line is that you make $58,000. If you think you can handle the payments after you retire, and if the question asked is "How much do you make now?", you're under no obligation, legal or moral, to mention your retirement in six months. Simply indicate your present income. You will be qualified on that basis. Remember, age cannot be a factor in granting your loan. An old rule of thumb was that your age plus the term of the loan could not total more that eighty years. We do not see that since recent federal laws outlaw discrimination based on age. An eighty-year-old man or woman can, in fact, obtain a thirty-year loan.

The lender will usually send a form to your employer, asking him to verify your income. If you're self-employed the lender will require copies of your last two years' income tax returns.

2. Some Other Tips For Filling Out Your Credit Application

On the back of the credit application there is an assets column. In this column you're asked the cash value of your insurance. The cash value of your insurance is not the face amount of your insurance. You might have $100,000 in insurance, but that is the policy value. The cash value on a term policy will be nil. If it's not a term policy, call your insurance agent to find out how much cash value it has.

Be careful what you put down as the value of various possessions. One of the biggest jokes in the credit office comes when someone pulls out a credit application and sees the value you have put on your 1978 Chevrolet. If you value your car at $3,000 and it's worth $300, they will question the rest of your application. Put a realistic value on your car and your land.

A piece of property you bought yesterday for $50,000 is probably worth $50,000 today. Some people buy a parcel of land in Dismal Acres for $50,000 and claim that it's worth $70,000. If you purchased a piece of property within the last two years for $50,000 it is probably worth $50,000. If it's worth more than that, let the lender find this out for himself.

Furniture and personal property should not be overvalued either. Don't try to impress the lender with assets such as furniture, antique jewelry, furs, coin collections, or gun collections. Minimize them.

Car payments are very important. If you're buying a $40,000 BMW, the payments could be $1,000 a month, which is more than many people pay on their home mortgage. The monthly payment on a $40,000 car will take up a substantial part of that "36" in the 28/36 underwriting ratio discussed earlier. So put off buying a car if you're getting ready to buy a home. And if you're really strapped for money, sell your fancy car and drive an old one until after your loan has been approved. We have seen people drive themselves out of a home loan by owning an expensive car with high monthly payments.

We've seen alimony hurt many loan applications. If you're separated or divorcing, don't lie about the alimony. The divorce decree will be a required part of your loan package if you are the one making the payment. It can also be part of your loan package if you're receiving alimony and want to rely on the alimony to qualify for the loan.

NOTE

...., 19.........   ...., Virginia
                                                                            [City]

....
[Property Address]

1. BORROWER'S PROMISE TO PAY

In return for a loan that I have received, I promise to pay U.S. $.. (this amount is called "principal"), plus interest, to the order of the Lender. The Lender is ....
.... I understand that the Lender may transfer this Note. The Lender or anyone who takes this Note by transfer and who is entitled to receive payments under this Note is called the "Note Holder."

2. INTEREST

Interest will be charged on unpaid principal until the full amount of principal has been paid. I will pay interest at a yearly rate of ....%.

The interest rate required by this Section 2 is the rate I will pay both before and after any default described in Section 6(B) of this Note.

3. PAYMENTS

(A) Time and Place of Payments

I will pay principal and interest by making payments every month.

I will make my monthly payments on the .... day of each month beginning on ....,
19......... I will make these payments every month until I have paid all of the principal and interest and any other charges described below that I may owe under this Note. My monthly payments will be applied to interest before principal. If, on
...., ..., I still owe amounts under this Note, I will pay those amounts in full on that date, which is called the "maturity date."

I will make my monthly payments at ....
.... or at a different place if required by the Note Holder.

(B) Amount of Monthly Payments

My monthly payment will be in the amount of U.S. $..

4. BORROWER'S RIGHT TO PREPAY

I have the right to make payments of principal at any time before they are due. A payment of principal only is known as a "prepayment." When I make a prepayment, I will tell the Note Holder in writing that I am doing so.

I may make a full prepayment or partial prepayments without paying any prepayment charge. The Note Holder will use all of my prepayments to reduce the amount of principal that I owe under this Note. If I make a partial prepayment, there will be no changes in the due date or in the amount of my monthly payment unless the Note Holder agrees in writing to those changes.

5. LOAN CHARGES

If a law, which applies to this loan and which sets maximum loan charges, is finally interpreted so that the interest or other loan charges collected or to be collected in connection with this loan exceed the permitted limits, then: (i) any such loan charge shall be reduced by the amount necessary to reduce the charge to the permitted limit; and (ii) any sums already collected from me which exceeded permitted limits will be refunded to me. The Note Holder may choose to make this refund by reducing the principal I owe under this Note or by making a direct payment to me. If a refund reduces principal, the reduction will be treated as a partial prepayment.

6. BORROWER'S FAILURE TO PAY AS REQUIRED

(A) Late Charge for Overdue Payments

If the Note Holder has not received the full amount of any monthly payment by the end of .... calendar days after the date it is due, I will pay a late charge to the Note Holder. The amount of the charge will be ....% of my overdue payment of principal and interest. I will pay this late charge promptly but only once on each late payment.

(B) Default

If I do not pay the full amount of each monthly payment on the date it is due, I will be in default.

(C) Notice of Default

If I am in default, the Note Holder may send me a written notice telling me that if I do not pay the overdue amount by a certain date, the Note Holder may require me to pay immediately the full amount of principal which has not been paid and all the interest that I owe on that amount. That date must be at least 30 days after the date on which the notice is delivered or mailed to me.

(D) No Waiver By Note Holder

Even if, at a time when I am in default, the Note Holder does not require me to pay immediately in full as described above, the Note Holder will still have the right to do so if I am in default at a later time.

(E) Payment of Note Holder's Costs and Expenses

If the Note Holder has required me to pay immediately in full as described above, the Note Holder will have the right to be paid back by me for all of its costs and expenses in enforcing this Note to the extent not prohibited by applicable law. Those expenses include, for example, reasonable attorneys' fees.

7. GIVING OF NOTICES

Unless applicable law requires a different method, any notice that must be given to me under this Note will be given by delivering it or by mailing it by first class mail to me at the Property Address above or at a different address if I give the Note Holder a notice of my different address.

Any notice that must be given to the Note Holder under this Note will be given by mailing it by first class mail to the Note Holder at the address stated in Section 3(A) above or at a different address if I am given a notice of that different address.

8. OBLIGATIONS OF PERSONS UNDER THIS NOTE

If more than one person signs this Note, each person is fully and personally obligated to keep all of the promises made in this Note, including the promise to pay the full amount owed. Any person who is a guarantor, surety or endorser of this Note is also obligated to do these things. Any person who takes over these obligations, including the obligations of a guarantor, surety or endorser of this Note, is also obligated to keep all of the promises made in this Note. The Note Holder may enforce its rights under this Note against each person individually or against all of us together. This means that any one of us may be required to pay all of the amounts owed under this Note.

9. WAIVERS

I and any other person who has obligations under this Note waive the rights of presentment and notice of dishonor and waive the homestead exemption. "Presentment" means the right to require the Note Holder to demand payment of amounts due. "Notice of dishonor" means the right to require the Note Holder to give notice to other persons that amounts due have not been paid.

10. UNIFORM SECURED NOTE

This Note is a uniform instrument with limited variations in some jurisdictions. In addition to the protections given to the Note Holder under this Note, a Mortgage, Deed of Trust or Security Deed (the "Security Instrument"), dated the same date as this Note, protects the Note Holder from possible losses which might result if I do not keep the promises which I make in this Note. That Security Instrument describes how and under what conditions I may be required to make immediate payment in full of all amounts I owe under this Note. Some of those conditions are described as follows:

Transfer of the Property or a Beneficial Interest in Borrower. If all or any part of the Property or any interest in it is sold or transferred (or if a beneficial interest in Borrower is sold or transferred and Borrower is not a natural person) without Lender's prior written consent, Lender may, at its option, require immediate payment in full of all sums secured by this Security Instrument. However, this option shall not be exercised by Lender if exercise is prohibited by federal law as of the date of this Security Instrument.

If Lender exercises this option, Lender shall give Borrower notice of acceleration. The notice shall provide a period of not less than 30 days from the date the notice is delivered or mailed within which Borrower must pay all sums secured by this Security Instrument. If Borrower fails to pay these sums prior to the expiration of this period, Lender may invoke any remedies permitted by this Security Instrument without further notice or demand on Borrower.

WITNESS THE HAND(S) AND SEAL(S) OF THE UNDERSIGNED.

....(Seal)
-Borrower

....(Seal)
-Borrower

....(Seal)
-Borrower

[Sign Original Only]

This is to certify that this is the Note described in and secured by a Deed of Trust dated ....
........................, 19.......... on the Property located in ...., Virginia.

My Commission expires:

....
Notary Public

VIRGINIA FIXED RATE NOTE—Single Family—FNMA/FHLMC UNIFORM INSTRUMENT   Form 3247 12/83

The lender will also look at rental income differently than you do. They don't care about your tax advantage as they're concerned with cash. If you have a negative cash flow on a rental property, it will probably work against you. But even if you have a positive cash flow, most lenders will only allow you to claim seventy-five percent of it as income. Why? Because there are going to be times when your property will be vacant.

We cannot emphasize too much the importance of minimizing your credit-card debt. A few years ago you could go to a bank and tell them not to worry about your credit-card debts, because you could secure a loan sufficient to build a house and pay off the credit cards. But the banks have learned that people who pay off huge credit-card debts are very likely to build them up again very quickly. So the lenders look for a pattern of going into debt. The best way to avoid this is to pay off your credit cards and other debts before making application for a loan. You can even wait a few months before seeking the loan if you have to. But do it. If you cannot, then make every effort to minimize your monthly debt payments in relation to the amount of the debt. If you have two credit-card debts and feel you can only afford to liquidate one, pay off the one in which the payment is highest in relation to the total debt. One might be ten percent of the debt and another might be three percent. Pay off the ten percent payment.

This is very important because the secondary lenders—Fannie Mae and others—will not buy a loan from a primary lender in which the borrower exceeds the 28/36 underwriting ratio. We have seen Freddie Mac refuse a loan with a 28/36.01 ratio! This would definitely apply if your credit cards show a pattern of indebtedness. The secondary lenders have to assume that this is an extension of your life-style, and that sooner or later you'll be financially overextended.

Finally, the manner in which you package your loan application and present your case is of the utmost importance. The president of a bank once told us that as a matter of personal and banking policies, he would never make a loan on a trailer in a particular subdivision. He would make a loan on a trailer or make a loan in this particular subdivision, but if both were involved he would decline because there were too many associated problems. But one day a young man walked into his bank and presented him with an application to put a trailer in that subdivision. The application was so complete, so overwhelming, and so convincing that he immediately approved the loan. So come properly prepared and properly budgeted. It does make a difference.

6 Insurance and More

When you're building and preparing to live in a house, there are four kinds of insurance with which you need to be concerned: (1) title insurance, (2) fire and EC insurance, (3) workmen's compensation insurance, and (4) general liability.

All of us understand the basic purpose of insurance and the protection it offers us. But not everyone understands that before we can obtain insurance for ourselves or our property we must have an insurable interest in the property, title, or person.

Insurable interest is an interest in a person or property to which injury or loss would adversely affect the insured party. For example, if your neighbor's house burns down you'd be sad, but you wouldn't suffer financially from the loss. Thus you don't have an insurable interest in your neighbor's house. But if you had loaned your neighbor money secured by a second trust on his house, you would have an insurable interest and would therefore be entitled to insure yourself against the loss of your neighbor's house.

What about the lender who has made the first loan trust on your neighbor's house? He also has an insurable interest and can insure himself against its loss.

What about life insurance on your neighbor or friend? If he dies you'll suffer no financial loss, only emotional loss. So you have no insurable interest in this life. But what if your neighbor is also your business partner? You would then have insurable interest because his death would have an impact on you financially.

In order to have a need for insurance, then, you must have an insurable interest. However, the decision to obtain insurance on your property or someone's life is usually a voluntary one.

A. TITLE INSURANCE

Title insurance protects the insured against the loss of his property because of a clouded title or no title (ownership) at all. When you take title to real property, as evidenced by the execution and delivery of a deed, it is logical to assume that the person deeding it to you owned the property free and clear of all liens and encumbrances. However, when you receive your deed you may not actually have a good title. Title insurance protects you against loss of ownership, not by guaranteeing that you retain the property, but by reimbursing you monetarily for the loss.

1. Owner's Title Insurance

The most important title insurance for you is owner's title insurance. This insurance protects you against a loss of title for the property.

2. Lender's or Mortgagee Title Insurance

The lender who makes you a loan secured by your property is called the mortgagee. Since the lender has an insurable interest in your property he will require a lender (mortgagee) title insurance policy for the amount of his loan.

Don't confuse owner's and lender's title insurance. The insured property is the same but there are two distinct insurable interests. Any lender will require lender's title insurance but owner's title insurance is optional. If the title to your home is invalid the lender would be paid his policy amount for the loss of his collateral, but if you elected not to purchase owner's title insurance you would have no insurance coverage and no home.

The lender will require that you obtain title insurance for him and that you pay for it. Whether you purchase owner's title insurance is your choice. Another confusing point is the policy amount. If you build a $100,000 home and only borrow $75,000 to build it you have a $100,000 insurable interest (your potential loss would be $100,000), but the lender's insurable interest is only $75,000. In this case you would be required to provide $75,000 in lender's title insurance and would choose whether you wanted to obtain the $100,000 in owners's title insurance. We should point out that the lender could not obtain more than $75,000 or you more than $100,000 since the insurable interest amount cannot exceed the potential loss. As your property value increases, however, you can and should increase your policy coverage to cover your potential loss.

3. What Does Title Insurance Protect?

A real estate purchase is one of the best investments you can make—so be certain to protect your land ownership against possible title problems that can hinder the transfer of your property or make it unmarketable. These problems are defects that occur before the date of the policy and remain undisclosed until sometime later. Even the most thorough search of the public records cannot reveal some of these "hidden" hazards.

A one-time premium safeguards your property from actual loss resulting from any risk covered by your policy, up to the amount of the policy, and provides for defense costs also, unless specifically excluded. Title insurance covers title defects such as:

- Forged deeds, mortgages, satisfaction of releases of mortgages and other instruments.
- False impersonation of the true owner of the land or of his consort.
- Instruments executed under fabricated or expired power of attorney (death or insanity of principal).
- Deeds apparently valid but actually delivered after death of grantor or grantee, or without consent of grantor.
- Deeds by persons of unsound mind.
- Deeds of minors.
- Deeds not properly delivered.
- Deeds which appear to convey title but are really mortgages.
- Outstanding prescriptive rights not of record and not disclosed by survey.
- Descriptions apparently but not actually adequate.
- Duress in execution of instruments.
- Defective acknowledgement due to lack of authority of notary. (Acknowledgement taken before commission or after expiration of commission.)
- Deed of property recited to be separate property of grantor which is in fact community or joint property.
- Deeds by persons apparently single but actually married.
- Deed from bigamous couple—prior existing marriage in another jurisdiction.
- Undisclosed divorce of spouse who conveys as sole heir of deceased consort.
- Undisclosed heirs.
- Misinterpretation of wills, deeds, and other instruments.
- Birth or adoption of children after date of will.
- Children living at date of will but not mentioned therein.
- Discovery of will or apparent intestate.
- Discovery of later will after probate of first will.
- Administration of estates and probate of wills of persons absent but not deceased.

- Conveyance by heir, devisee, or survivor of a joint estate who murdered the decedent.
- Deed from trustees of purported business trust which is in fact a partnership or joint stock association.
- Deed of executor under nonintervention will when order of solvency has been fraudulently procured or entered.
- Deeds to or from corporations before incorporation or after surrender of forfeiture of charter.
- Claims of creditors against property conveyed by heirs or devisees within prescribed period after owner's death.
- Mistakes in recording legal documents. (For example, incorrect indexing, errors and omissions in transcribing, and failure to preserve original instruments.)
- Record easement, but erroneous ancient location of pipe or sewer line which does not follow route of granted easement.
- Special assessments where they become lien upon passage of resolution and before recordation or commencement of improvements for which assessed.
- Want of jurisdiction of persons in judicial proceedings.
- Failure to include necessary parties in judicial proceedings.
- Federal estate and gift tax liens.
- State inheritance and gift tax liens.
- Errors in tax records. (For example, listing payment against wrong property.)
- Ineffective waiver of tax liens by tax or other governing authorities repudiated later by successors.
- Corporation franchise taxes as lien on all corporate assets, notice of which does not have to be recorded in the local recording office.
- Erroneous reports furnished by tax officials, but not binding on municipality.
- Tax homestead exemptions set aside as fraudulently claimed.
- Lack of capacity of foreign personal representatives and trustees to act.
- Deeds from nonexistent entities.
- Interests arising by deeds to fictitious characters to conceal illegal activities on the premises.
- Deeds in lieu of foreclosure set aside as being given under duress.
- *Ultra vires* deed given under falsified corporate resolution.
- Conveyances and proceedings affecting rights of servicemen protected by Soldiers and Sailors Civil Relief Act.
- Federal condemnation without filing of notice. (Federal law does not require filing of notice of taking in local recording office.)

- Break in chain of title beyond period of examination of public records where running of adverse possession statute has been suspended. (True owner is incompetent, absent, or incarcerated or title is held by the sovereign.)
- Deed from record owner of land where he has sold property to another purchaser on unrecorded land contract and the purchaser has taken possession of premises.
- Void conveyances in violation of public policy. (Payment of gambling debt, payment for contract to commit crime, or conveyance made in restraint of trade.)
  (Reprinted with permission from the American Title Insurance Company, San Francisco, California.)

Those fifty things are most of the problems that might affect the ownership of your property. As a rule if something goes wrong the lawyer who did the title search cannot be held responsible. He has a professional responsibility to certify the title. But if he searches the title, examines the documents, and decides they are in order, it's not his responsibility to attest to the validity of the signatures, improper filings, or other defects in the title. If in a divorce case, the spouse says, "I never signed that deed," is the lawyer responsible? Can you sue him? No. Do you own your property? No.

Even when the lawyer does his job you can be in trouble. The possibility also exists that the lawyer did not do his job, or that he was fraudulent. Can you go back to the lawyer and get a judgment? Possibly, but the lawyer may be judgment-proof. A judgment against him could be uncollectable. Lawyers may perform the title work for an entire subdivision, and probably do not have enough malpractice insurance to cover all of it. Or there could be error or fraud which would deny you the title. You can't rely on a lawyer to insure that you have a good title.

4. The Cost

In round figures the first title insurance policy for you or the lender is going to cost about $2.50 per $1,000. The premium for the same coverage for both you and the lender would be $4.25 if you buy them together. In other words, you get a discount on the second policy. The purchase of title insurance is a one-time-only cost which is good for as long as you own your property, but it is not transferable. You can, however, increase your coverage as your house appreciates. When you sell the property the new owner must obtain new title coverage for himself. He may, however, be able to negotiate a lower rate if he buys insurance from the same company which had insured the former owner.

5. Affirmative Mechanics Lien Coverage

In some states, if a contractor or subcontractor files a mechanic's lien during the building of a house, and it is ruled valid, this lien will take precedence over your lender's lien. This is called piercing the deed of trust. If a mechanic works on your home and you do not pay him, he can file a lien and ultimately sell your property out from under the lender in order to satisfy the lien. The lender, therefore, requires you to secure additional title insurance protection and affirmative mechanic's lien coverage. This policy protects the lender, not the borrower.

The affirmative mechanic's lien covereage applies only during the construction process. The owner, not the builder, must obtain it for the lender, and it is the lender who benefits if the lienor is due the money. Affirmative mechanic's lien coverage is only required in some states.

6. Recommendation

Paying the premium is only part of the insurance process. There are instances where buyers pay the premiums but never receive the policy. The policy may not be issued because of a title problem, an administrative error, or because the insurance company did not actually receive the premium. So insist on physical possession of your insurance policy.

The policy itself is only as valuable as the coverage it specifically provides. Some policies provide protection subject to certain exceptions. If your policy has exceptions discuss them with your lawyer.

B. HOMEOWNERS INSURANCE

Unlike title insurance, homeowner's insurance is a single policy shared by you and your lender. In a homeowner's policy you are listed as the owner and the lender is listed as the "loss payee." In the case of a loss, the lenders are paid first, with any excess (equity) going to the owner.

This insurance covers your house in the event loss due to fire, wind, hail, falling trees, and other causes. The latter items are termed extended coverage. The basic insurance on your home is protection against fire. All fire insurance policies are the same throughout the United States. They are patterned after the New York State Fire Insurance Policy and every paragraph in every policy by an insurer is identical.

1. Homeowner's Policy

A homeowner's policy, which covers fire and extended coverage, will also include other insurance coverage, the most important of which is general liability. For example, if a man comes to your house to deliver a letter, trips on the sidewalk and breaks his neck, he can sue you if you were negligent. The general liability coverage under your homeowner's policy would protect you against this suit.

Other items are covered by a homeowner's policy. Within the insurance industry, homeowner's policies are frequently called Homeowner's 1, Homeowner's 2, Homeowner's 3, and so forth. The larger the number or the fancier the title, the more the coverage. We recommend a Homeowners 3 or comparable. You'll pay a nominal additional amount—perhaps ten to fifteen percent more for a Homeowners 3 policy—but it will provide broad liability coverage, protect the contents of your house, cover items stolen from your car, and so on. Of course it also provides normal fire and extended coverage.

2. Flood Insurance

When you secure a building permit, the county and your lender will want to know if you are located in a floodplain. If you are, you will have to provide the lender with flood insurance. Homeowner policies do not protect you from flooding. Flood insurance must be obtained from the Federal Insurance program. It is not difficult to obtain, but it is not included in your other policies and is not standard.

3. Builder's Risk Policy

A builder's risk policy is a misnomer. It does not insure you against a poor builder or a careless builder who might set off a fire. It is not a policy the builder procures in order to protect you. It is a homeowner's policy that is only applicable during the construction period. It gives you the same coverage as a homeowner's policy, but takes into consideration the fact that your house is under construction.

When your home is finished, a builder's risk policy becomes a homeowner's policy. If your insurance company writes you a policy for the construction period and calls it a homeowner's policy, you are protected. Be sure to tell them, however, that the house is still under construction.

One small point to remember about a builder's risk policy that might save you a little money: You don't require any insurance until you have something insurable, such as a foundation. Even a foundation would be difficult to damage by fire. If you had purchased a typical builder's risk or homeowner's policy at settlement, you and your lender might go a month or more without any insurable interest in the project except for liability. Accordingly you may elect not to purchase insurance until you have actually made improvements to your property.

4. Riders

When you obtain a builder's risk policy, it probably will not include a policy for theft or vandalism. These are specific riders which you have to request. They are very expensive and sometimes very difficult to obtain. They have different provisions as theft is different from vandalism. If you want this protection, ask about the cost and decide accordingly.

5. Co-insurance

When you buy a $20,000 piece of property and subsequently build an $80,000 home on it your insurable interest amount is $80,000. Your land (except for trees) cannot burn down. You and your lender can obtain insurance for up to $80,000. But what if there is no lender and you decide that you want only $50,000 in insurance? This will create a problem because the law requires you to obtain fire and extended coverage insurance for at least eighty percent of your insurable interest. This is called the co-insurance clause. If you don't maintain insurance for at least eighty percent of your insurable interest then you are a co-insurer with your insurance company.

To illustrate, assume you own a $20,000 lot and $80,000 home for an $80,000 insurable interest. If you carry insurance in the amount of $80,000 and the house is fifty percent destroyed by fire, then your insurance company will pay you $40,000. If your house is totally destroyed then you would receive $80,000.

But if you had bought only $40,000 in insurance (fifty percent of your insurable interest), you would be a co-insurer with your insurance company, because your policy is less than eighty percent of your insurable interest. In case of a $40,000 fire loss, you would only receive $20,000, not $40,000. As a co-insurer you and the insurer split the loss.

To be protected, you must have coverage of at least eighty percent of your insurable interest. To receive the full $40,000 of your loss in the above case amount, you would need $64,000 worth of coverage because that is eighty percent of your insurable interest.

The co-insurance provision is rarely a problem. But it can be a serious problem for you if you underinsure your home. Review your insurance coverage periodically and/or maintain an inflation rider in your policy to make sure you are properly protected.

C. WORKER'S COMPENSATION

Worker's compensation is mandated by Congress and administered by the states. If an employee is injured on the job, he cannot file suit against his employer if the employer has worker's compensation. The state worker's compensation board reviews the case and determines the amount of compensation to be paid to the claimant-employee. The law has removed liability from the employer (the contractor) provided that he has worker's compensation insurance. The system is designed to make sure that all injured parties are compensated and employers are kept out of the courtroom. But if workers are not covered by worker's compensation insurance, they could conceivably sue you if injured on the job. So it is critical that all contractors on your job maintain worker's compensation insurance. Your lender will frequently require evidence of this coverage.

Insist that all contractors provide you with a certificate that they have worker's compensation insurance. This is a reasonable requirement, and critical to the financial safety of your project.

D. GENERAL LIABILITY

As previously discussed, general liability insurance protection protects you from risks other than fire and hail losses. Your general liability coverage will probably not protect you or your contractor if he does something wrong. You should require your contractor to provide a certificate of general liability coverage as well as his worker's compensation certificate.

For example, what if a contractor working on your house had a fight with his wife or girlfriend? He's angry as a hornet, and he decides to leave your property on a driveway that hasn't been built yet. He literally goes straight through your house and a neighbor's house as well. Is that covered by worker's compensation? No—but it would be covered by a general liability policy.

To repeat, your contractor should show you evidence (certificates) of *both* a General Liability Policy and Worker's Compensation. Two different risks, two different policies, and two separate requirements that you as prudent homebuilder should make sure are fulfilled.

7 The Appraisal

The appraisal presents a potential problem for someone building a custom home. Obtaining an accurate appraisal of a condominium or tract house built by a large developer is easy since hundreds of them have been built and sold in the same neighborhood. It is easy for the appraiser to put a reliable market value on such a house. Most appraisers use "comparables" as their primary measuring rod in placing a value on a home. They find similar houses in similar neighborhoods that have sold recently, and then assume that your house would sell for around the same amount. But if you have a unique house, it is virtually impossible to find a "comparable" for it. Adding up the cost of the land, material, and labor obviously does not give the true value of a home. The result sometimes can be so ridiculous that the so-called "science" of appraising is a misnomer.

A. THE SCIENCE OF APPRAISING

If appraising is a science, it is a very subjective one. Appraisers try to present their services as the product of an objective industry, but this isn't entirely accurate.

There are three methods of determining value by appraisal: the income approach, the cost approach, and the market or comparable value approach.

The income approach determines value by computing a capitalized value based on the amount of rental income the property can produce. As a rule, the income approach produces a value significantly less than the other methods for owner-occupied homes and is often not used.

The cost approach is the most technical and perhaps the most nearly objective. But it does not necessarily represent true value. Here the appraiser actually determines the cost for each component, including the land, well, septic, driveway, basement, floor, wall, roof, fireplace, and so on. In theory the collective cost of these items should represent the project value but frequently it does not. In fact, some projects cost more than they are worth, while other projects are worth significantly more than they cost. So the cost approach is not very reliable. Even so, it is important to make a detailed record of the cost of your house.

The most widely used and accepted method of appraisal is the comparable value method.

Under the comparable approach the appraiser will identify certain characteristics of your home such as style and size. He then will find comparable homes in your area and the sale price of these homes. He then will make adjustments based on lot size and other special features to determine the value of your home. If other comparable homes are selling for $100,000 in your neighborhood, then your home must be worth $100,000.

Unfortunately this approach does not always work well for a custom home because it frequently has no close comparables.

The appraiser will do his best to find comparables, but we have found that a custom home is compared with homes of comparable size. If this happens the comparable approach will undervalue your project. You may not be able to do much about it (we offer some suggestions later) but you need to be prepared for this possibility.

B. COMPARABLE CONSIDERATIONS

1. The Six-Month Rule

One element you'll contend with in the comparable approach is the six-month rule. In order for a home to be used as a comparable, it must have been sold within the last six months. There may be another home worth $100,000 in a $75,000 neighborhood but if it has not sold then its cost can't be used as its value.

2. Comparable Style

Another element to consider is the comparable style. If you are building a contemporary home in a predominantly traditional neighborhood the appraiser will have trouble finding comparables. Home styles different from those in the neighborhood frequently appraise for less.

3. Rural Location

What if you plan to build a "custom home" in a remote location? Comparables aren't very easy to find, which frequently presents a problem.

4. The Unique Home

The same problem arises if you build a very unique home. In this case it is quite possible that you'll be unable to obtain a comparative appraisal because there are *no* comparables. Without a satisfactory appraisal it will be difficult to obtain a normal loan, and you may have to obtain a portfolio loan.

5. The Thirty-Percent Rule

There is a general appraisal rule that the value of the land, as a percentage of the total project, can't be greater than thirty percent (and for some lenders no more than twenty-five percent). If the land value exceeds thirty percent then your project is not acceptable.

For example, if you have a lot worth $50,000, and plan to build a home worth $200,000, your total project cost would be $250,000, and the land cost would represent only twenty percent of the project value, which presents no problem. However, if you wanted to build a $100,000 home on the same lot you would have a problem because the total project cost is $150,000 and the land cost is thirty-three percent of the total. Since the land cost is greater than thirty percent, under the thirty-percent rule this would be an unacceptable loan.

The thirty-percent rule is not ironclad, however, and there are numerous exceptions. Still, it can be a formidable problem in areas where land values are unusually high.

C. THE NEED FOR APPRAISALS

If appraisals are so unscientific and even misleading, then why do you need one? Lenders demand appraisals because the law requires them. Even if the law didn't require them, however, most lenders would insist on them as a matter of policy. It is a form of lender protection, and gives them something to fall back on if a requested loan seems too large or one with which they are uncomfortable.

A home also can be appraised before it is built. The appraiser looks at your land, your plans and specifications, and estimates what your home will be worth when it is finished. A sample appraisal report is included at the end of this chapter.

D. TIPS ON OBTAINING A GOOD APPRAISAL

Obtaining a good (proper) appraisal may be critical to your project, so it's important that you prepare for the appraisal process. Here's a checklist of

Property Description & Analysis   **UNIFORM RESIDENTIAL APPRAISAL REPORT**   File No. _____

SUBJECT

Property Address			
City	County	State	Census Tract
			Zip Code
Legal Description			Map Reference
Owner/Occupant			
Sale Price $	Date of Sale		
Loan charges/concessions to be paid by seller $			
R.E. Taxes $	Tax Year	HOA $/Mo.	
Lender/Client			

LENDER DISCRETIONARY USE

Sale Price $ _____
Date _____
Mortgage Amount $ _____
Mortgage Type _____
Discount Points and Other Concessions
Paid by Seller $ _____
Source _____

PROPERTY RIGHTS APPRAISED
☐ Fee Simple
☐ Leasehold
☐ Condominium (HUD/VA)
☐ De Minimis PUD

NEIGHBORHOOD

LOCATION	☐ Urban	☐ Suburban	☐ Rural
BUILT UP	☐ Over 75%	☐ 25-75%	☐ Under 25%
GROWTH RATE	☐ Rapid	☐ Stable	☐ Slow
PROPERTY VALUES	☐ Increasing	☐ Stable	☐ Declining
DEMAND/SUPPLY	☐ Shortage	☐ In Balance	☐ Over Supply
MARKETING TIME	☐ Under 3 Mos.	☐ 3-6 Mos.	☐ Over 6 Mos.

PRESENT LAND USE	%	LAND USE CHANGE	PREDOMINANT	SINGLE FAMILY HOUSING		
			OCCUPANCY	PRICE $ (000)	AGE (yrs)	
Single Family	___	☐ Not Likely	☐ Owner			
2-4 Family	___	☐ Likely	☐ Tenant	Low		
Multi-family	___	☐ In process	☐ Vacant (0-5%)	High		
Commercial	___	To: ___	☐ Vacant (over 5%)	Predominant		
Industrial	___			–		
Vacant	___					

NEIGHBORHOOD ANALYSIS

	Good	Avg.	Fair	Poor
Employment Stability	☐	☐	☐	☐
Convenience to Employment	☐	☐	☐	☐
Convenience to Shopping	☐	☐	☐	☐
Convenience to Schools	☐	☐	☐	☐
Adequacy of Public Transportation	☐	☐	☐	☐
Recreation Facilities	☐	☐	☐	☐
Adequacy of Utilities	☐	☐	☐	☐
Property Compatibility	☐	☐	☐	☐
Protection from Detrimental Cond.	☐	☐	☐	☐
Police & Fire Protection	☐	☐	☐	☐
General Appearance of Properties	☐	☐	☐	☐
Appeal to Market	☐	☐	☐	☐

Note: Race or the racial composition of the neighborhood are not considered reliable appraisal factors.

COMMENTS: _____

SITE

Dimensions			
Site Area		Corner Lot	
Zoning Classification		Zoning Compliance	
HIGHEST & BEST USE: Present Use		Other Use	

UTILITIES	Public	Other		SITE IMPROVEMENTS	Type	Public	Private
Electricity	☐	☐		Street		☐	☐
Gas	☐	☐		Curb/Gutter		☐	☐
Water	☐	☐		Sidewalk		☐	☐
Sanitary Sewer	☐	☐		Street Lights		☐	☐
Storm Sewer	☐	☐		Alley		☐	☐

Topography	
Size	
Shape	
Drainage	
View	
Landscaping	
Driveway	
Apparent Easements	
FEMA Flood Hazard	Yes* ☐ No ☐
FEMA* Map/Zone	

COMMENTS (Apparent adverse easements, encroachments, special assessments, slide areas, etc.): _____

IMPROVEMENTS

Stories	Exterior Walls
Type (Det./Att.)	Roof Surface
Design (Style)	Gutters & Dwnspts.
Existing	Window Type
Proposed	Storm Sash
Under Construction	Screens
Age (Yrs.)	Manufactured House
Effective Age (Yrs.)	

	Crawl Space		% Finished	
	Basement		Ceiling	☐ Ceiling
	Sump Pump		Walls	☐ Walls
	Dampness		Floor	☐ Floor
	Settlement		Outside Entry	☐ None
	Infestation			Adequacy
				Energy Efficient Items:

ROOM LIST

ROOMS	Foyer	Living	Dining	Kitchen	Den	Family Rm.	Rec. Rm.	Bedrooms	# Baths	Laundry	Other	Area Sq. Ft.
Basement												
Level 1												
Level 2												

Finished area **above** grade contains: Rooms; Bedroom(s); Bath(s); Square Feet of Gross Living Area

INTERIOR

SURFACES	Materials/Condition
Floors	
Walls	
Trim/Finish	
Bath Floor	
Bath Wainscot	
Doors	

HEATING		COOLING	
Type		Central	
Fuel		Other	
Condition		Condition	
Adequacy		Adequacy	

KITCHEN EQUIP.		ATTIC	
Refrigerator ☐		None ☐	
Range/Oven ☐		Stairs ☐	
Disposal ☐		Drop Stair ☐	
Dishwasher ☐		Scuttle ☐	
Fan/Hood ☐		Floor ☐	
Compactor ☐		Heated ☐	
Washer/Dryer ☐		Finished ☐	
Microwave ☐			
Intercom ☐			

Fireplace(s) #

IMPROVEMENT ANALYSIS	Good	Avg.	Fair	Poor
Quality of Construction	☐	☐	☐	☐
Condition of Improvements	☐	☐	☐	☐
Room Sizes/Layout	☐	☐	☐	☐
Closets and Storage	☐	☐	☐	☐
Energy Efficiency	☐	☐	☐	☐
Plumbing-Adequacy & Condition	☐	☐	☐	☐
Electrical-Adequacy & Condition	☐	☐	☐	☐
Kitchen Cabinets-Adequacy & Cond.	☐	☐	☐	☐
Compatibility to Neighborhood	☐	☐	☐	☐
Appeal & Marketability	☐	☐	☐	☐
Estimated Remaining Economic Life				Yrs.
Estimated Remaining Physical Life				Yrs.

AUTOS

CAR STORAGE:	
Garage ☐	Attached ☐
Carport ☐	Detached ☐
None ☐	Built-In ☐

Adequate ☐	House Entry ☐
Inadequate ☐	Outside Entry ☐
Electric Door ☐	Basement Entry ☐

COMMENTS

Additional features:

Depreciation (Physical, functional and external inadequacies, repairs needed, modernization, etc.):

General market conditions and prevalence and impact in subject/market area regarding loan discounts, interest buydowns and concessions:

Freddie Mac Form 70 10/86 **12Ch.**   Forms and Worms Inc.® 315 Whitney Ave., New Haven, CT 06511   1(800) 243-4545   Item #130960   Fannie Mae Form 1004 10/86

Valuation Section   **UNIFORM RESIDENTIAL APPRAISAL REPORT**   File No.

Purpose of Appraisal is to estimate Market Value as defined in the Certification & Statement of Limiting Conditions.

COST APPROACH

BUILDING SKETCH (SHOW GROSS LIVING AREA ABOVE GRADE)

If for Freddie Mac or Fannie Mae, show only square foot calculations and cost approach comments in this space.

ESTIMATED REPRODUCTION COST – NEW – OF IMPROVEMENTS:

Dwelling _____ Sq. Ft. @ $ _____ = $ _____
_____ Sq. Ft. @ $ _____ = $ _____

Extras _____ = $ _____

Special Energy Efficient Items _____ = $ _____
Porches, Patios, etc. _____ = $ _____
Garage/Carport _____ Sq. Ft. @ $ _____ = $ _____
Total Estimated Cost New _____ = $ _____

Less   Physical | Functional | External
Depreciation _____ = $ _____
Depreciated Value of Improvements .... = $ _____
Site Imp. "as is" (driveway, landscaping, etc.) .. = $ _____
ESTIMATED SITE VALUE .... = $ _____
(If leasehold, show only leasehold value.)
INDICATED VALUE BY COST APPROACH .... = $ _____

(Not Required by Freddie Mac and Fannie Mae)

Does property conform to applicable HUD/VA property standards? ☐ Yes ☐ No

If No, explain:

Construction Warranty ☐ Yes ☐ No
Name of Warranty Program
Warranty Coverage Expires

The undersigned has recited three recent sales of properties most similar and proximate to subject and has considered these in the market analysis. The description includes a dollar adjustment, reflecting market reaction to those items of significant variation between the subject and comparable properties. If a significant item in the comparable property is superior to, or more favorable than, the subject property, a minus (–) adjustment is made, thus reducing the indicated value of subject; if a significant item in the comparable is inferior to, or less favorable than, the subject property, a plus (+) adjustment is made, thus increasing the indicated value of the subject.

ITEM	SUBJECT	COMPARABLE NO. 1		COMPARABLE NO. 2		COMPARABLE NO. 3	
Address							
Proximity to Subject							
Sales Price	$		$		$		$
Price/Gross Liv. Area	$	☑ $		☑ $		☑ $	
Data Source							
VALUE ADJUSTMENTS	DESCRIPTION	DESCRIPTION	+ (–) $ Adjustment	DESCRIPTION	+ (–) $ Adjustment	DESCRIPTION	+ (–) $ Adjustment
Sales or Financing Concessions							
Date of Sale/Time							
Location							
Site/View							
Design and Appeal							

LYSIS

SALES COMPARIS

	Total	Bdrms	Baths		Total	Bdrms	Baths		Total	Bdrms	Baths		Total	Bdrms	Baths
Above Grade Room Count															
Gross Living Area		Sq. Ft.				Sq. Ft.				Sq. Ft.				Sq. Ft.	
Basement & Finished Rooms Below Grade															
Functional Utility															
Heating/Cooling															
Garage/Carport															
Porches, Patio, Pools, etc.															
Special Energy Efficient Items															
Fireplace(s)															
Other (e.g. kitchen equip., remodeling)															
Net Adj. (total)	+		– $		+		– $		+		– $		+		– $
Indicated Value of Subject			$				$				$				$

Comments on Sales Comparison:

RECONCILIATION

INDICATED VALUE BY SALES COMPARISON APPROACH .... $

INDICATED VALUE BY INCOME APPROACH (If Applicable) Estimated Market Rent $ _____ /Mo. x Gross Rent Multiplier _____ = $

This appraisal is made [ ] "as is" [ ] subject to the repairs, alterations, inspections or conditions listed below [ ] completion per plans and specifications.

Comments and Conditions of Appraisal:

Final Reconciliation:

This appraisal is based upon the above requirements, the certification, contingent and limiting conditions, and Market Value definition that are stated in

[ ] FmHA, HUD &/or VA instructions.

[ ] Freddie Mac Form 439 (Rev. 7/86)/Fannie Mae Form 1004B (Rev. 7/86) filed with client _____ 19 _____ [ ] attached.

I (WE) ESTIMATE THE MARKET VALUE, AS DEFINED, OF THE SUBJECT PROPERTY AS OF _____ 19 _____ **to be $** _____

I (We) certify: that to the best of my (our) knowledge and belief the facts and data used herein are true and correct; that I (we) personally inspected the subject property, both inside and out, and have made an exterior inspection of all comparable sales cited in this report; and that I (we) have no undisclosed interest, present or prospective therein.

Appraiser(s) SIGNATURE _____   Review Appraiser SIGNATURE _____   [ ] Did [ ] Did Not

NAME _____   (if applicable) NAME _____   Inspect Property

Freddie Mac Form 70 10/86 12Ch. MC3/89 Forms and Worms Inc.® 315 Whitney Ave. New Haven, CT 06511 1(800) 243-4545 **Item #** 130960 Fannie Mae Form 1004 10/86

Property Description & Analysis — **UNIFORM RESIDENTIAL APPRAISAL REPORT** — File No. ___

SUBJECT

Property Address	
City	County	State	Census Tract	Zip Code
Legal Description		Map Reference	
Owner/Occupant		
Sale Price $	Date of Sale	
Loan charges/concessions to be paid by seller $	
R.E. Taxes $	Tax Year	HOA $/Mo.	
Lender/Client		Source	

PROPERTY RIGHTS APPRAISED
- Fee Simple
- Leasehold
- Condominium (HUD/VA)
- De Minimis PUD

LENDER DISCRETIONARY USE

Sale Price	$
Date	
Mortgage Amount	$
Mortgage Type	
Discount Points and Other Concessions	
Paid by Seller	$

NEIGHBORHOOD

LOCATION: Urban / Suburban / Rural
BUILT UP: Over 75% / 25-75% / Under 25%
GROWTH RATE: Rapid / Stable / Slow
PROPERTY VALUES: Increasing / Stable / Declining
DEMAND/SUPPLY: Shortage / In Balance / Over Supply
MARKETING TIME: Under 3 Mos. / 3-6 Mos. / Over 6 Mos.

PRESENT LAND USE %
- Single Family ___
- 2-4 Family ___
- Multi-family ___
- Commercial ___
- Industrial ___
- Vacant ___

LAND USE CHANGE
- Not Likely
- Likely
- In process
- To: ___

PREDOMINANT OCCUPANCY
- Owner
- Tenant
- Vacant (0-5%)
- Vacant (over 5%)

SINGLE FAMILY HOUSING
PRICE $ (000)	AGE (yrs)
Low	
High	
Predominant —	

NEIGHBORHOOD ANALYSIS

	Good	Avg.	Fair	Poor
Employment Stability				
Convenience to Employment				
Convenience to Shopping				
Convenience to Schools				
Adequacy of Public Transportation				
Recreation Facilities				
Adequacy of Utilities				
Property Compatibility				
Protection from Detrimental Cond.				
Police & Fire Protection				
General Appearance of Properties				
Appeal to Market				

Note: Race or the racial composition of the neighborhood are not considered reliable appraisal factors.

COMMENTS: _____

SITE

Dimensions	
Site Area	Corner Lot
Zoning Classification	Zoning Compliance
HIGHEST & BEST USE: Present Use	Other Use

UTILITIES	Public	Other	SITE IMPROVEMENTS	Type	Public	Private
Electricity			Street			
Gas			Curb/Gutter			
Water			Sidewalk			
Sanitary Sewer			Street Lights			
Storm Sewer			Alley			

Topography	
Size	
Shape	
Drainage	
View	
Landscaping	
Driveway	
Apparent Easements	
FEMA Flood Hazard	Yes* / No
FEMA* Map/Zone	

COMMENTS (Apparent adverse easements, encroachments, special assessments, slide areas, etc.):

IMPROVEMENTS

Units
Stories
Type (Det./Att.)
Design (Style)
Existing
Proposed
Under Construction
Age (Yrs.)
Effective Age (Yrs.)

Foundation
Exterior Walls
Roof Surface
Gutters & Dwnspts.
Window Type
Storm Sash
Screens
Manufactured House

Slab
Crawl Space
Basement
Sump Pump
Dampness
Settlement
Infestation

Area Sq. Ft.
% Finished
Ceiling
Walls
Floor
Outside Entry

Roof
Ceiling
Walls
Floor
None
Adequacy
Energy Efficient Items:

Area Sq. Ft.

ROOM LIST

ROOMS	Foyer	Living	Dining	Kitchen	Den	Family Rm.	Rec. Rm.	Bedrooms	# Baths	Laundry	Other	Area Sq. Ft.
Basement												
Level 1												
Level 2												

Finished area **above** grade contains: Rooms; Bedroom(s); Bath(s);

Square Feet of Gross Living Area

INTERIOR

SURFACES — Materials/Condition
Floors
Walls
Trim/Finish
Bath Floor
Bath Wainscot
Doors

Fireplace(s) #

HEATING
Type
Fuel
Condition
Adequacy
COOLING
Central
Other
Condition
Adequacy

KITCHEN EQUIP.
Refrigerator
Range/Oven
Disposal
Dishwasher
Fan/Hood
Compactor
Washer/Dryer
Microwave
Intercom

ATTIC
None
Stairs
Drop Stair
Scuttle
Floor
Heated
Finished

IMPROVEMENT ANALYSIS
Quality of Construction
Condition of Improvements
Room Sizes/Layout
Closets and Storage
Energy Efficiency
Plumbing-Adequacy & Condition
Electrical-Adequacy & Condition
Kitchen Cabinets-Adequacy & Cond.
Compatibility to Neighborhood
Appeal & Marketability
Estimated Remaining Economic Life ___ Yrs.
Estimated Remaining Physical Life ___ Yrs.

Good Avg. Fair Poor

AUTOS

CAR STORAGE:
Garage
Carport
None

No. Cars ___
Condition ___

Attached
Detached
Built-In

Adequate
Inadequate
Electric Door

House Entry
Outside Entry
Basement Entry

Additional features:

COMMENTS

Depreciation (Physical, functional and external inadequacies, repairs needed, modernization, etc.):

General market conditions and prevalence and impact in subject/market area regarding loan discounts, interest buydowns and concessions:

Freddie Mac Form 70 10/86 **12Ch.**   Forms and Worms Inc.® 315 Whitney Ave., New Haven, CT 06511 · 1(800) 243-4545   Item #130960   Fannie Mae Form 1004 10/86 things to do:

(1) Provide the appraiser with a legal description of your property (either a copy of the deed or the purchase agreement).

(2) If possible, provide a site plan (a proposed house location survey is preferred).

(3) Provide good directions to your property, including a map and a narrative. We cannot emphasize how much this will help the appraiser, and how much he will appreciate it.

(4) Provide detailed plans and specifications. They give the impression of a well-organized project and a more sophisticated approach, which could result in a higher appraisal. The more detailed your plans and specifications, the easier for the appraiser to visualize the final project.

(5) If possible, provide an artistic rendering of the home to be built. A picture is worth a thousand words.

And finally, remember how important comparables are in the appraisal process. Try to locate comparables for your appraiser. If you are able to provide your appraiser with good comparables you should be well on your way to a fair appraisal.

We recommend strongly that you talk to your appraiser. In your conversation you should help with directions and comparables. This demonstrates to the appraiser that you have an interest and concern in your project. The appraiser could be making three or four appraisals in one day, or fifteen or twenty in a single week. If you call him about your appraisal you establish personal contact. Keep it to a two-to-three minute phone conversation. Let him know you understand that he is appraising your home and you would like to point out some special features about your home that might help him with his appraisal. Also discuss your warranty. The more comprehensive the warranty, the more value that can be applied to your home.

E. LOW APPRAISALS

Do not be discouraged if your appraisal comes in lower than expected. It is your personal project and you should try to make it work. First, discuss the problem with the appraiser. Go over each item to make sure he included everything in his appraisal. Items frequently omitted are basement space, fireplaces, decks, garages, and special upgrades such as cedar shingles and $2'' \times 6''$ walls. Also, check the comparables and offer new ones that support a higher appraisal.

If that doesn't work, hire another appraiser. In most cases this will be acceptable to the lender. Another appraiser may see your project differently. Before you hire another appraiser check with the lender to confirm that obtaining another appraisal is acceptable to them and that they approve the new appraiser you plan to employ.

8 How to Select and Negotiate with Contractors

Selecting and negotiating with contractors and subcontractors is the essence of building a house. A full-time builder does not have to do any actual construction work himself. He contracts it out to a variety of skilled tradesmen—plumbers, electricians, carpenters, roofers, drywall installers and finishers, painters, floor men, and carpet installers. The people he chooses to build his house and the prices he negotiates with them will ultimately determine not only the quality of the home but also its cost, which will have a bearing on his profit.

The people you pick to build your house will determine the quality of your home and the cost of that house. Finding the right person at the right price is essential. You don't want hasty, shoddy work, but you don't want to pay too much or your cost of construction will be far out of line with other houses in the marketplace.

Most people who decide they don't want to build their own home are uncomfortable with contractors and subcontractors. They don't know which one is competent, whether the job is being done right, or what the fair price is.

These are all legitimate concerns. This chapter is designed to guide you through the process of selecting and negotiating with the tradesmen who will actually build your house. The process can be exasperating at times but it isn't as difficult as most people think. Men and women from all walks of like have done it, and most people who do come away with a great sense of satisfaction—not only in having built their own home, but demonstrating their ability to deal with people and make sensible, cost-effective decisions.

A. WHERE TO START

1. The Yellow Pages

As good as place as any to start looking for a subcontractor is in the yellow pages. As the saying goes, a person who fails to advertise, advertises to fail. Generally speaking, a person who advertises is successful. So don't reject a subcontractor as probably too costly just because he happens to have the largest ad in the yellow pages. You may well wish to select him because he has the largest ad—if he can meet all your needs.

2. Friends

Another way to start looking is to ask friends if they know any subcontractors. Rarely do builders or subcontractors ever get praised. Keep asking questions in order to find out just how good he really is and whether he is dependable. The fact is that most subcontractors—and the more successful they are, the more likely they are to do it—schedule more work than they can handle. They simply find it hard to turn down an attractive job. So they take on more work and then try to juggle their schedules and keep everyone happy. However, the busy, over-committed subcontractor is usually a pretty good craftsman, even though he may have done something to irritate your friend. Find out what bothered your friend. It may turn out to be something inconsequential which you can live with if the subcontractor's work is good and is reasonably priced.

When you ask a friend about a subcontractor, and he starts yelling that the man in question threw trash on the ground or was late, he still may be a good subcontractor. The fact is that only very seldom is anyone going to praise a subcontractor no matter how good he might be. So sift through the trivial and find out whether the man did his work properly, and whether it was good work. If it was, he is the subcontractor you ought to consider.

3. The Better Business Bureau

The Better Business Bureau is not the best place to look for subcontractors, because if your contractor is a bad one, he will definitely be a member of the Better Business Bureau. If he is a good contractor he will probably also be a member of the Better Business Bureau, but if he is a bad one, we guarantee he will be. So when you walk into a contractor's office and he has a plaque on the wall stating that he is a member of the Better Business Bureau, the only thing to be deduced from that is that he's a member.

4. Building Material Companies

Building material companies are excellent sources of builders and subcontractors. Very likely they won't recommend a contractor or subcontractor who is not good. Even though a subcontractor buys material from them they will not recommend a bad one. They might remain silent or noncommittal, but more than likely they will recommend or suggest only contractors they know to be competent and reliable. Building material suppliers are in the front line of the building industry. Their information is usually good, as they have a reputation to maintain.

5. Neighborhood Job Sites

A job site in your neighborhood where someone is building a home is a good place to look for subcontractors. You will find working contractors there. You can ask about their work and will probably have an opportunity to see them actually working on the job and to assess their performance.

B. WHERE TO CONFIRM

There are other ways to check out a contractor.

1. Other Contractors

A good way to confirm a contractor's work is through another contractor. But if you're asking about a plumber, do not ask another plumber. Ask an electrician or a carpenter. One plumber will rarely recommend another one, nor will an electrician recommend another. But sometimes you can learn a lot about a plumber by talking to an electrician who has worked with him and vice versa.

2. Homebuilder's Association

If a contractor is a member of the local homebuilder's association, this is usually a good sign that he has some interest in his work and what is going on in the building trade. It is not inexpensive to belong to the homebuilder's association. The good, successful builders are usually members. If a contractor or subcontractor is not doing his job—building in accordance with code and keeping his people happy—he can be expelled from the association.

3. The Contractor's Bank

This may surprise you, but there is a very high correlation between a plumber's ability to plumb and his bank account. The better plumbers have better bank accounts and enjoy good standing with their bankers. If a plumber is a reputable and honest worker, he also has a good bank account—not necessarily a big bank account, but one that is handled in a way that suggests he manages his affairs well. Under the privacy law a bank is not permitted to divulge information about one of its customers without the customer's written permission. So we suggest that you ask your plumber for permission to talk with his banker. If the plumber says yes, that alone should make you feel confident about him. If the banker says that he is a good customer it probably means that he is a good plumber as well. That applies to all the other persons involved in building your house as well.

4. Quality of Work

One of the best ways to assess a subcontractor's work performance is to look at his work. This isn't always easy to arrange, but it can be done. Seeing his actual work—especially if you know something about that kind of work—will certainly help you to assess his skills.

5. Appearance and Car

How a plumber or an electrician looks or dresses or what kind of car he drives will tell you absolutely nothing. He could be dressed properly and be a terrible plumber, or vice versa. He could drive a large car and be a poor electrician, or drive a wreck of a van and be an excellent electrician.

6. Office

An office can be an important reflection of a subcontractor's status. The office might be in his home, which is fine. But if he will be doing several thousand dollars worth of work on your house you want to know whether he has an office and how easy he is to reach (see below).

7. Licensed, Bonded and Insured

As we discussed previously, the fact that the subcontractor has the phrase, "licensed, bonded, and insured" printed on the side of his van or pickup truck means very little. Almost all subcontractors are licensed, bonded, and insured. The real question is—For how much and for what?

8. Contractor's License

A contractor's license is very important. If he is a plumber he should have a plumber's license, an electrician should have an electrician's license, and so on. There are different types of builders' licenses. There may be a class A or class B license. If a person is going to work on your house in a particular trade and your state requires that trade to be licensed, then you want to confirm that the tradesman has the proper license.

Sometimes a contractor will come out and plumb your house. You pay him and when you go for the inspection the lender asks, "Who plumbed your house?"

"Johnny Jones," you say.

"He's not licensed," the building inspector says. "We won't inspect."

If the state requires a particular tradesman to have a license, be sure that he does.

9. Business License

This one is not important, since it's not going to affect his ability to do the work on your home.

10. Telephone and Phone Check

A telephone number is more important than you might think. If a subcontractor gives you his telephone number, call information or look in the telephone book to see if his number is listed. It is better to do business with a subcontractor whose number is in the phone book. Although it may be that for personal reasons he doesn't want his name in the phone book, if his name is not listed, it is more likely that he will not be a very good plumber.

It is a good idea to call the subcontractor and leave a message when he isn't there. How long does it take him to return your call? If a subcontractor cannot find the time to return a phone message, it's likely that he can't plumb your house within the time frame you'd like.

C. NEGOTIATIONS

The first thing you need to know is that in proper negotiations both parties come out ahead. Do not hesitate to negotiate. Negotiation is a way of life in some cultures. You can enroll in classes in negotiations anywhere in the country. There is a great deal involved in pricing a product, and in theory anything is negotiable.

1. How to Secure the Best Bid At The Best Price

Do not hesitate to negotiate with a subcontractor. In all probability he will expect it. But don't bid against yourself. This means that if you are selling a car and someone says, "How much are you asking for it?", you should not name a price. It is very hard to do, but if you give them a price, you are bidding against yourself, because the purchaser will then counter with a lower figure.

Never be the one to quote the first price. To translate this principle to the building business, don't suggest that you are looking for a bid in the range of $3,000. If you do, the subcontractor will probably bid around $2,900 when he may have been willing to do the work for $2,500.

2. The Process

Solicit several bids—a minimum of three, but it is probably a good idea to get five. It will take a little more effort to obtain five bids, but if one of those bids saves you $300 to $500, it will be worth it. Most of the subcontractors you are soliciting won't have the sophistication to submit accurate bids, they frequently will be guessing. Since everything is negotiable you'll want an assortment of bids.

But don't then throw out the highest and lowest bids. Some people follow the theory that the low bidder doesn't know what he is doing and the high bidder is looking for a windfall. Look at all the bids carefully.

When you're soliciting bids, inform all the subcontractors that you are securing several quotations as it puts a little more pressure on them. Also tell them you are building soon—for example, "Please quote plumbing on the house to be plumbed in August." This will make a difference since the subcontractor will know you are serious and will reply more quickly.

Then say, "Rush the bid." This also suggests that you mean business. Or specify a date by which you want the bid submitted. If the contractor misses the date, eliminate him. He is the same contractor who can't return your phone call. However, if he calls and says, "I can't get it by such-and-such a date, but I'll have it in by such-and-such," that's another matter. Put a star by him. He is a good manager.

3. When the Bids Come In

When the bids come in, select those that seem to be what you are looking for in terms of price, service, and professionalism. Inform the finalists that their bids have come close. This tells the contractor several things: (1) his bid is being considered; (2) you're serious, a person who is looking for services; (3) you're still thinking about it and haven't made up your mind yet; and (4), the subcontractor may be just a few dollars away from getting the contract. Do not elaborate on that. Do not tell him how much the other bidders are quoting. Do not tell him who the other bidders are, although he probably knows.

There are a couple of things you can do after the bids come in, especially when you're dealing with a subcontractor who will bid on a job for you only once. Let's assume that you like his bid, but he doesn't know you do. So he calls and says, "How about my bid?"

Your response should be, "Wait a minute. Let me look." Then you say, "Ouch!" Or, if you are talking in person and he gives you the bid, you respond with what is known as the flinch. Your gesture should not be offensive, although some contractors (especially if they have done their homework and feel they have given you a good bid) might say, "Buzz off, bumpkin head!" This, however, is unlikely. It is entirely possible that your flinch will prompt him to refigure his bid. In the process he could discover that he charged you too much, and lower his quotation. The worst that could happen is that he will defend his bid which is what you want. You are looking for a bidder who can discuss and defend his bid intelligently. It will make both of you feel comfortable.

4. Unbundling

At this point you want to focus a little more sharply on the bid and try to narrow it down to its components. This is called unbundling the bid, and even experienced bidders hate the process. When he bids, you say, "Oh thank you, Mr. Plumber. Now would you mind breaking the bid down into labor, material, profit, and overhead?"

A subcontractor does not have to unbundle his bid, and if he does not do it, you will have to make your decision based on the composite bid. Do not be too concerned if he does not unbundle. If he does unbundle it may be a potential bonanza for you; the two of you can then reevaluate the pricing, with the result that you both may benefit.

5. The Disguised Discount

Sometimes negotiators get involved in what we call the disguised discount, which works like this: A builder gives you a bid of $50,000 for the shell of a home. The shell is the floor, walls, roof, siding, windows, doors, and partitions. You give him the flinch. Then you ask him to break down his quotation.

He says, "I'm not unbundling. You can flinch all you want, but that's a good price."

On the other hand, the contractor may want the business and may have given you a fair bid. But at this stage of the negotiation he is afraid to reduce the price because he will lose face and credibility. So give him the opportunity to reevaluate the $50,000 price and reduce it without losing face. So now ask, "What can you take out or change to lower the price?"

At this point, there are two things the contractor can do. He can calculate the cost differential between the Andersen windows he has included and less expensive windows and give you a new bid. You gain nothing by this. You can get the house you want for $50,000 or less of a home for the actual window cost difference.

But there is another possible scenario. The bidder also has a chance to reduce his original price indirectly. If his cost differential between Andersen and the less expensive windows is $3,000, he may reduce his price by $4,000 if you chose the less expensive window. Since his cost differential is only $3,000 he has actually reduced his total bid by $1,000. We call this a disguised discount.

At this point you can accept the lower price and the generic windows or can suggest that you will supply the Andersen windows and take the home for $46,000 less his cost of the generic windows. Rather than go through all of the hassle, the bidder at this point will probably say: "Oh, I'll give you the whole house with the Andersen windows for $49,000.

By suggesting a reworking of the specifications and the bid, you gave the bidder an indirect way of reducing his price (a disguised discount) by giving a $4,000 reduction for something that only reduced his cost by $3,000.

6. What Is the Best Price?

In all probability, subcontractors will not give you their best price at first. Contractors do not like to be asked for their best price, so what they generally do on a best price bid is to give a price but also be a little vague on what it includes. This leaves the door open.

7. Compromise Gradually

If the subcontractor is asking $10,000 and you are saying $5,000 and he proposes to split the difference, what do you do?

You don't split the difference. You know he is willing to go down to $7,500, so begin bargaining between his $7,500 and your $5,000.

If he says he wants $10,000 and you say you want $5,000, and he comes down to $7,500, you offer $5,200. If he then says, "No, I need more" and you first go to $5,400 and then to $5,600, you are showing him what to expect the next time: $5,800. What you should do is to go to $5,200, $5,300, $5,350, $5,375, $5,380, and so on—a little at a time. Give up only gradually.

8. Do Not Squeeze

Finally, do not squeeze your contractor too hard trying to get that final, best price. If he needs the work and he has given you what he thinks is a fair price, but you press him to come down another $50 or $100, he might agree. But then he may make it up by doing shoddy or hasty work so that he can get on to the next job where he thinks the price is fair.

9 A Sequence for Building Your Home

There are many ways to build a home. Builders have learned from experience that in some cases certain things must be done before others. Construction frequently goes much more smoothly if you do one thing first and then another. If you build more than one home you may develop some sequencing ideas of your own. But when building your first home it will help to draw on the experience of others. The method we present here is not the only way to build a home, but it has been developed by a number of experienced people, including our colleague Peter Hotz, who is an architect and has designed and built many custom homes.

A. THE DRAW SCHEDULE

It is important that you establish a satisfactory draw schedule at the time you negotiate your construction loan (see Chapter 5), because the draw schedule will determine the sequence of your construction work. Most lenders have had much experience with residential loans and their draw schedule will be properly sequenced. Nevertheless, it's a good idea to know how many draws you want and what you want to accomplish with each draw even though the lender may propose a different schedule. It is possible that your lender may be willing to work with your schedule instead. This chapter is based on the draw schedule in Chapter 5.

The schedule is divided into five segments or phases, and you will be able to see how they relate to the draw schedule. With slight modifications our sequencing can also be used with other schedules with more segments.

B. SUBCONTRACTORS AND SUPPLIERS

It is necessary to know which subcontractors and suppliers you will need to complete your house. We have provided a list of subcontractors you will probably need to hire and a description of the work they do. Following that is another list of contractors, with space for you to insert the name of the successful bidder you have selected to do the work. You may not need all of them on your project or you may need to add some specialty trades or suppliers. Also included is a detailed construction schedule which is based on the experience of the construction industry. They have a logical sequence and suggest the activities you will be dealing with in the preliminary phases—surveying, portable toilets, trash removal; move on through concrete masonry and metals for the foundation; then into woods (carpentry), insulation and moisture protection (roofing) for the shell; and on into the finish trades.

LIST OF SUBCONTRACTORS YOU WILL NEED

- *Surveyor:* stakes the house foundation, sets the building's corners (or "points") in the footings and prepares the wall check and final survey.
- *Excavator:* digs the foundation and footings, installs the sub-drainage system and backfills and rough grades the site.
- *Paver:* installs the asphalt driveway.
- *Landscaper:* installs retaining walls, finishes the grading, seeding or sodding and planting.
- *Concrete Supplier:* delivers ready-mixed concrete to the construction site.
- *Mason:* installs the foundation wall and brick or stone veneer (if any). He also constructs masonry fireplace and masonry landscape items such as walls, barbecues, and terraces. If natural stone is used in masonry work, it will need to be obtained from a quarry. If the mason does not supply his own materials, such things as structural steel beams, angles, and decorative iron railings must be ordered from various suppliers.
- *Carpenter:* erects the wood frame of the house, installs the interior partition framing, sheathing and wood siding, exterior windows and trims, all doors, cabinetry and countertops, builds fixed bookcases, installs closet shelves and poles, stairs, railings and other miscellaneous millwork.
- *Lumber supplier:* furnishes all material installed by the carpenter and frequently can furnish other products, such as metal fireplaces, insulation, drywall, electrical and plumbing fixtures, kitchen cabinets,

and bathroom vanities. Stairs may sometimes be obtained from the lumber supplier or from a separate millwork company.

- *Waterproofer:* supplies and installs the liquid asphalt material for the below-ground foundation walls.
- *Insulation:* supplies and installs batt insulation in the floors (if needed) and blanket or blown insulation in the attic. Rigid insulation on the exterior walls is normally supplied by the lumber subcontractor and installed by the carpenter.
- *Roofer:* supplies and installs the shingles or other roofing material and any rigid insulation used in the roof.
- *Weatherstripper:* insulates doors, although this is often done by the carpenter.
- *Sheet metal:* supplier will cut and form flashing to desired shapes.
- *Glazer:* furnishes and installs special shapes of fixed glass not available from window suppliers—such as triangles, trapezoids, and half-round fixed windows.
- *Windows:* can be obtained from window manufacturers or the lumber supplier. Usually the same supplier will also stock a line of sliding glass doors.
- *Doors:* exterior and interior doors are usually obtained from the lumber supplier, but special wood doors can be ordered from door manufacturers or millwork shops. The hardware is obtained from a locksmith or lumber supplier.
- *Drywall:* this subcontractor furnishes, installs, and finishes gypsum drywall on the interior walls and ceilings. Although it is rarely done today, plaster is still an alternative to drywall if you can find and afford the skilled subcontractor.
- *Flooring:* material for floors include ceramic tile, vinyl, carpeting and hardwood, although the carpeting and wood flooring is usually installed by separate contractors. It is, however, possible to find one subcontractor who will install any kind of flooring.
- *Painting and Wallpapering:* These subcontractors may or may not furnish their own material. The quality of paint and wallpaper varies considerably, so it is often preferable to select material from a supplier and have it installed by a separate subcontractor.
- *Bathroom Accessories:* can be supplied by a specialty company or by the lumber company and installed by the carpenter.
- *Kitchen Cabinets:* kitchen cabinets as well as bathroom vanities can be obtained from specialty suppliers or from the lumber supplier; countertops can be supplied by the same supplier or specialty company. All cabinets are installed by the carpenter.
- *Appliances:* are available from various retail and discount outlets

or from a cabinet supplier. They are installed by the supplier, a plumber, or an electrician.

- *Plumbing:* one subcontractor normally installs all the pipe and fixtures. The plumbing fixtures can be obtained from a special supplier or the lumber company.
- *Heating—Ventilating—Air-conditioning (HVAC):* one subcontractor supplies and installs all the equipment and ductwork necessary to heat, ventilate, and cool the house. If the heating system uses hot water, the plumber may supply and install the system.
- *Electrician:* supplies and installs all electrical wiring, power panel, outlets, and switches. Lighting fixtures are obtained from an electrical supplier and installed by the electrician.

YOUR SUBCONTRACTORS

1. General Requirements:

    Surveying:

    Temporary Facilities:

    Trash Removal:

2. Site Work

    Excavating Sub:

    Paving Sub:

    Landscape Sub:

3. Concrete

    Concrete Sub:

    Concrete:

4. Masonry

    Masonry Sub:

5. Metals

    Metals:

6. Wood & Plastics

    Carpentry Sub:

    Lumber:

    Stairs:

7. Thermal & Moisture Protection

    Waterproofing Sub:

    Insulation Sub:

    Roofing Sub:

    Weatherstripping Sub:

    Flashing:

8. Doors & Windows

    Garage Door Sub:

    Fixed Glass & Glazing:

    Windows:

    Sliding Glass Doors &

    Interior Doors:

    Entrance Doors:

    Door Hardware:

9. Finishes

    Drywall Sub:

    Flooring Sub:

    Hardwood Sub:

    Painting & Wallpaper Sub:

    Paint & Wallpaper:

10. Specialties

Bath Accessories:

11. Casework

Cabinets:

Countertops:

Appliances:

12. Mechanical

Plumbing Sub:

HVAC Sub:

13. Electrical

Electrical Sub:

Electrical Fixtures:

C. THE CRITICAL PATH METHOD (CPM)

There are many ways to chart the sequence in the building a house. Most builders eventually develop a method of their own, based on experience and what seems to work best for them. One of the methods we recommend is called the critical path method (CPM) schedule. The schedule presented in this book was adapted to the construction of a single-family residence by Peter Hotz.

The CPM schedule, originally developed between 1956 and 1975 for planning construction, is a graphic model of a construction project. Our CPM schedule is divided into five phases. To see how it works, look at Phase 1. This corresponds to the first schedule, for the first 20% of the house. It starts at number 1 with staking the corners (physically locating the house on the site) and ends with framing the first floor deck at number 13. The CPM schedule continues through the next four phases of construction corresponding to the draw schedule. The CPM schedule can be used as the only sequencing aid to building your home. However, through our use of it we have found that another method of scheduling based on the CPM but with certain differences was better.

D. THE CONSTRUCTION SCHEDULE CHECKLIST

One of the problems in relying solely on the CPM schedule is that it does not take into account *lead time*—the amount of time it takes to get certain trades or materials to the job site. Therefore, we developed a checklist that will keep you on the critical path, but that will also alert you to items that could take more than a week to get to the job. It can be very disconcerting when you realize that you should have ordered some material a month ago, and that it is not in stock when you need it.

We recommend that you work closely with the Construction Schedule Checklist presented here. It is an effective, easy way of scheduling your work and following through on it. Although this is a good comprehensive list of the activities that have to be done and the correct sequence in which they are done, there may be some things that are unique to your construction project which you have to add to your checklist.

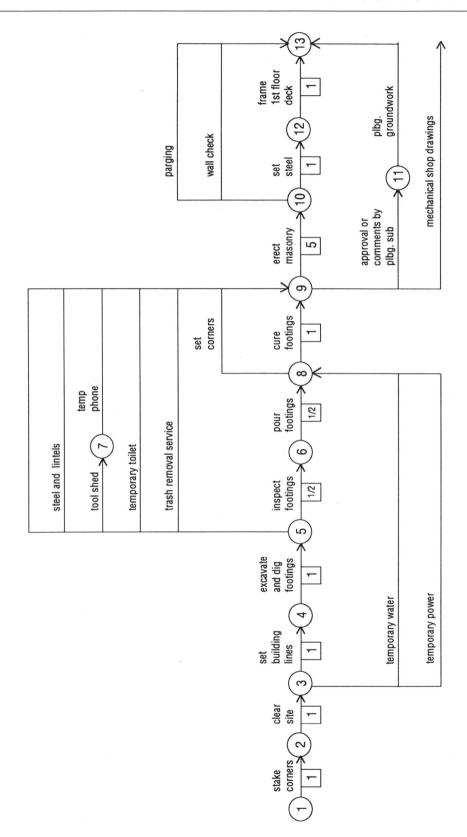

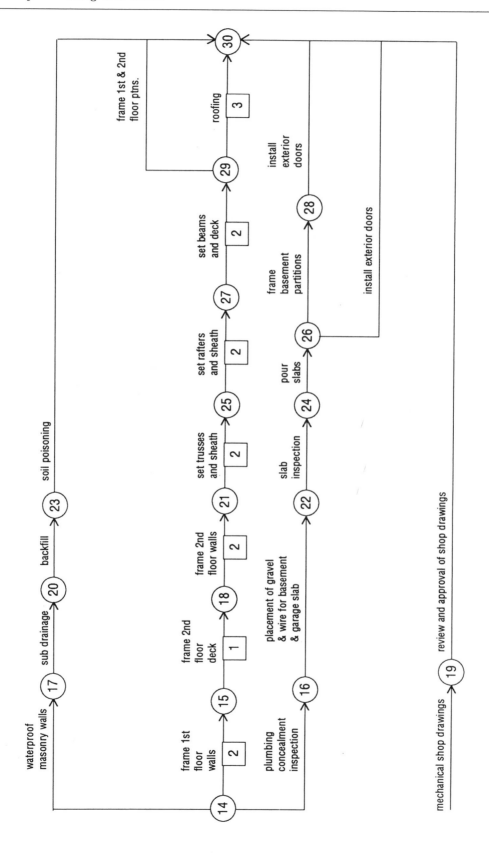

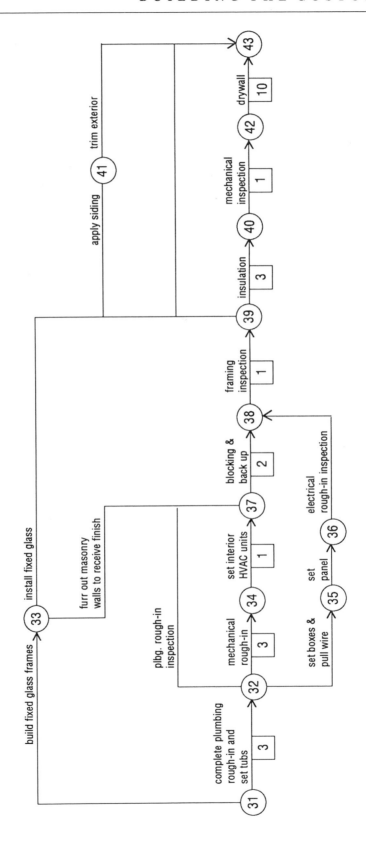

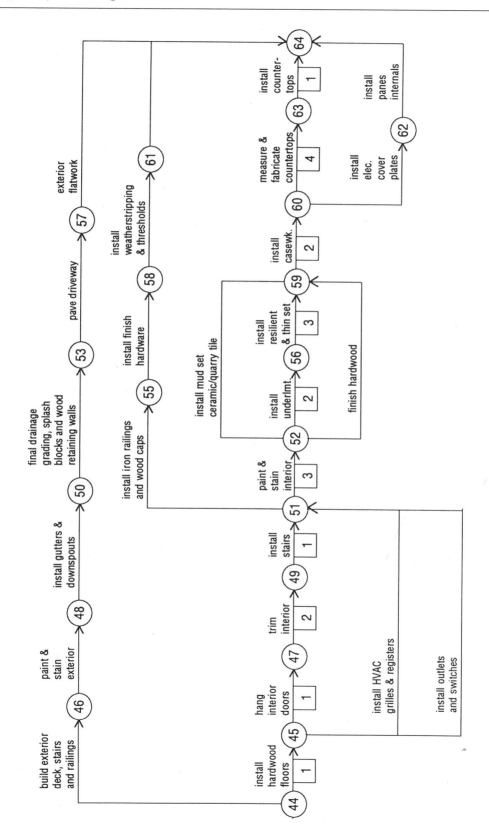

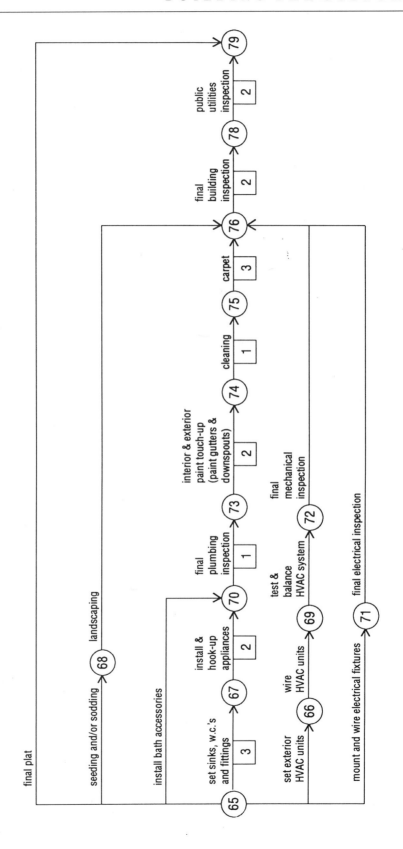

YOUR CONSTRUCTION SCHEDULE CHECKLIST
Preliminary Phase

A
- ☐ order masonry set windows for delivery in 3 weeks (1A)
- ☐ obtain building permit

B
- ☐ request surveyor to stake offsets in 1 week (C)
- ☐ contract with lumber supplier
- ☐ contract for excavating to start in 2 weeks (1A)
- ☐ order 1st floor trusses for delivery in 4 weeks (1C)

C
- ☐ schedule excavating and rough grading
- ☐ schedule digging of footings
- ☐ schedule installation of temporary power
- ☐ schedule installation of temporary water

- ☐ order structural steel for delivery in 2 weeks (1B)
- ☐ order balance of windows for delivery in 3 weeks (1C)
- ☐ contract for masonry to start in 2 weeks (1B)
- ☐ order roof trusses for delivery in 5 weeks (2B)

Phase One

1A
- ☐ house location staked
- ☐ building site cleared and excavated
- ☐ footings marked and dug
- ☐ footings inspected and poured
- ☐ temporary water, power, toilet, phone
- ☐ trash removal

- ☐ schedule mason
- ☐ check on steel delivery

- ☐ contract with carpenter to start framing in 2 weeks (1C)
- ☐ contract with plumber to install groundwork in 2 weeks (1C)

☐ order bathtubs/showers for installation in 6 weeks (3A)
☐ order metal fireplaces for delivery in 4 weeks (2B)
☐ order skylights for delivery in 4 weeks (2B)

1B   ☐ masonry complete
☐ steel set

☐ order lumber to first floor deck
☐ schedule carpenter to start framing
☐ schedule plumber to do groundwork

☐ order special hardware for delivery in 12 weeks (3C)
☐ contract with roofer to start in 4 weeks (2C)
☐ request wall check from surveyor for next week (1C)

1C   ☐ wall check survey
☐ first floor deck framed
☐ plumbing groundwork
☐ request first construction loan advance

☐ order lumber to roof
☐ schedule waterproofing of basement walls
☐ order sub-drainage materials

☐ contract with exterminator to treat soil in 2 weeks (2B)
☐ contract with concrete finisher to pour basement and
garage in 2 weeks (2B)
☐ order casework for delivery in 12 weeks (4C)

Phase Two

2A   ☐ frame 1st floor walls and partitions
☐ waterproof masonry walls
☐ install sub-drainage
☐ backfill inspection
☐ plumbing concealment inspection

- [ ] order roof ventilators
- [ ] schedule backfilling
- [ ] order gravel, reinforcing wire & bar, and poly for garage & basement slabs
- [ ] schedule exterminator
- [ ] schedule concrete finisher

- [ ] order exterior doors for delivery in 2 weeks (2C)
- [ ] order garage door installation in 2 weeks (2C)
- [ ] contract with glass company to install fixed glass in 4 weeks (3B)
- [ ] order brick or stone veneer for delivery in 3 weeks (3A)

2B

- [ ] frame to roof & sheathe
- [ ] backfill
- [ ] finish grade, gravel, wire & poly for basement & garage slab
- [ ] soil poisoning
- [ ] slab inspection
- [ ] pour slab
- [ ] install metal fireplaces
- [ ] install skylights

- [ ] schedule roofer
- [ ] order lumber if required to frame basement partitions
- [ ] schedule delivery of exterior doors
- [ ] schedule installation of garage door

- [ ] contract with HVAC contractor to start rough-in in 3 weeks (3B)
- [ ] order special electrical fixtures for delivery in 12 weeks (5A)
- [ ] order recessed fixture and fan housings for delivery in 4 weeks (3C)

2C

- [ ] roofing
- [ ] frame basement partitions
- [ ] set exterior doors
- [ ] install garage door
- [ ] request second construction loan advance

- [ ] schedule plumber to complete rough-in and set tubs (3A)
- [ ] order material for fixed-glass frames (3A)

Phase Three

3A
- [ ] complete plumbing rough-in & set tubs
- [ ] plumbing rough-in inspection
- [ ] build fixed-glass frames
- [ ] exterior siding and/or veneer (1st of 3 possible weeks)

- [ ] schedule HVAC rough-in
- [ ] order furring strips for finished masonry walls
- [ ] schedule fixed-glass installation

- [ ] contract with electrical sub to start in 2 weeks (3C)

3B
- [ ] HVAC rough-in
- [ ] set interior HVAC units
- [ ] furr out all masonry walls to receive finish
- [ ] exterior siding and/or veneer (2nd of 3 possible weeks)
- [ ] install fixed glass

- [ ] schedule electrical rough-in
- [ ] order blocking and back-up material

- [ ] contract with insulator to start in 2 weeks (3D)
- [ ] contract with drywall contractor to start in 3 weeks (3E)

3C
- [ ] electrical rough-in & inspection
- [ ] blocking and back-up
- [ ] framing inspection
- [ ] exterior siding and/or veneer complete
- [ ] special hardware on site
- [ ] install recessed fixtures and fan housings

- [ ] schedule insulation
- [ ] order exterior trim materials

- [ ] order stairs for delivery in 4 weeks (4A)
- [ ] order plumbing fixtures and fittings for installation in 8 weeks (5A)

3D
- ☐ insulation
- ☐ HVAC inspection
- ☐ concealment checklist
- ☐ trim exterior (1st of 3 possible weeks)

- ☐ schedule drywall hanging
- ☐ provide for temporary heat if necessary

- ☐ contract for ceramic tile to be installed in 5 weeks (4C)
- ☐ order stock items of finish hardware for delivery in 4 weeks (4B)

3E
- ☐ hang drywall
- ☐ trim exterior (2nd of 3 possible weeks)

- ☐ schedule drywall finishing

- ☐ order hardwood for delivery in 2 weeks (4A)
- ☐ contract for hardwood installation to start in 2 weeks (4A)
- ☐ order stocked electrical fixtures for delivery in 6 weeks (5A)
- ☐ contract with carpet contractor to start in 8 weeks (5C)

3F
- ☐ finish drywall
- ☐ trim exterior
- ☐ request third construction loan advance

- ☐ schedule hardwood installation
- ☐ order interior trim lumber
- ☐ check delivery of stairs
- ☐ schedule installation of HVAC grilles and registers
- ☐ schedule installation of electrical outlets and switches

- ☐ request gutters and downspouts for installation in 2 weeks (4B)
- ☐ contract for paving to start in 3 weeks (4C)
- ☐ contract for resilient flooring to start in 3 weeks (4C)

Phase Four

4A
- ☐ install hardwood
- ☐ hang interior doors
- ☐ trim interior
- ☐ install stairs
- ☐ install HVAC grilles & registers
- ☐ install outlets and switches

- ☐ schedule painting of interior and exterior
- ☐ schedule installation of gutters & downspouts
- ☐ order exterior deck, stair, railing and trim material
- ☐ order stair and interior balcony railings and handrails
- ☐ order underlayment material
- ☐ check hardware availability

- ☐ contract with hardwood finisher to start in 2 weeks (4C)
- ☐ order bath accessories for delivery in 4 weeks (5A)

4B
- ☐ interior paint & stain
- ☐ build exterior decks, stairs, & railings
- ☐ complete all exterior trim
- ☐ exterior paint and stain
- ☐ gutters & downspouts installed
- ☐ install stair and interior balcony railings & handrails
- ☐ install finish hardware
- ☐ install underlayment

- ☐ order RR ties and re-bar
- ☐ schedule paving of driveway
- ☐ schedule installation of ceramic tile
- ☐ schedule installation of resilient flooring
- ☐ schedule finishing of hardwood
- ☐ schedule delivery of casework

- ☐ contract for weatherstripping and thresholds to be installed in 2 weeks (4D)
- ☐ order countertops for delivery in 2 weeks (4D)
- ☐ order appliances for delivery in 4 weeks (5B)

4C
- [ ] final drainage grading, install splashblocks & wood retaining walls
- [ ] pave driveway
- [ ] install ceramic and/or quarry tile flooring & tub surround
- [ ] install resilient flooring
- [ ] finish hardwood
- [ ] install casework

- [ ] schedule concrete finisher for exterior flatwork
- [ ] schedule installation of weatherstripping & thresholds
- [ ] schedule delivery of countertops
- [ ] schedule installation of electrical cover plates & panel internals

- [ ] contract for landscaping (seed, sod, mulch, planting) to start in 2 weeks (5A)

4D
- [ ] exterior flatwork
- [ ] install weatherstripping & thresholds
- [ ] install countertops
- [ ] install electrical cover plates
- [ ] install electrical panel internals
- [ ] request fourth construction loan advance

- [ ] schedule seeding and/or sodding and/or planting and/or delivery of mulch
- [ ] check availability of bath accessories and electrical fixtures
- [ ] schedule installation of sinks, w.c.'s, and fittings
- [ ] schedule installation of exterior HVAC equipment
- [ ] schedule installation of electrical fixtures

Phase Five

5A
- [ ] final plat by surveyor
- [ ] seeding, sodding & planting
- [ ] install bath accessories
- [ ] set sinks, w.c.'s, and fittings
- [ ] set & wire exterior HVAC units
- [ ] mount and wire electrical fixtures

☐ check delivery of appliances and other equipment
☐ schedule final hook-up and balancing of HVAC equipment
☐ request plumbing, mechanical, and electrical final inspections for next week

☐ request painting of gutters & downspouts, & touch-up in 2 weeks (5C)
☐ contract for cleaning to start in 2 weeks (5C)

5B
☐ install and hook up appliances & other equipment
☐ final plumbing inspection
☐ test and balance HVAC system
☐ final mechanical inspection
☐ final electrical inspection

☐ schedule painting of gutters & downspouts (if required) & touch-up
☐ schedule cleaning
☐ schedule carpet installation

5C
☐ interior & exterior paint touch-up
☐ paint gutters and downspouts
☐ clean interior
☐ clean windows
☐ install carpet

☐ check all work for completion
☐ schedule final building and public utilities inspections

☐ request final plat from surveyor

5D
☐ final building inspection
☐ public utilities inspection
☐ request fifth construction loan advance

Our checklist does not include site preparation—such things as the survey, well or septic system, lot clearing, or access to the property. These are done before construction begins on the house. Our checklist pertains strictly to the house construction.

The items in the checklist are identified by a number corresponding to the draw schedule or phase and a letter which suggests a one-week time frame. If you stay on this entire schedule, accomplishing each week as outlined, you will deserve a presidential citation, for you will have mastered the most difficult job of the construction manager—bringing all the materials and tradesmen to the job exactly when they are needed.

Realistically, our schedule is a target at which to aim. You will find that what is planned as a week's worth of work on the schedule may take two weeks. Time is important, and the schedule will constantly challenge you to keep your job moving toward completion.

When we use the word schedule in the checklist, we mean two or three telephone calls made to the contractor to make sure that he will be there when you want him. One of these calls should be made the night before he is scheduled to be present. Most contractors are hard to reach by telephone during the day, and often they do not come home until 6:00 or 7:00 P.M., so that's when you have to catch them. Be considerate and don't call them too late in the evening. They usually go to bed early because they get up early.

E. THE PRELIMINARY PHASE

The preliminary phase, the first page of our checklist, does not correspond to the draw schedule. These are lead-time items that need to be accomplished before construction begins. They are divided into groups A, B, and C, each representing a week.

If your house has a masonry or poured concrete basement with windows, the windows need to be ordered three weeks before construction will start. Also, obviously you will need a building permit which (depending upon where your house is to be built) may take a couple of weeks or even a couple of months.

It is a good idea to visit the county building department a month or two before you are ready to start construction and familiarize yourself with their requirements. If you have your construction drawings (building departments usually require two sets for permits) you can apply for a permit while you are there—provided that you are confident about building in the near future. In this way you are sure to obtain your permit about three weeks before the start of construction.

Two weeks before construction is to begin, schedule your surveyor to locate the house with offset stakes. These are usually placed ten feet away from each corner of the house, so they will not be disturbed during excavation. You or your carpentry contractor should have a complete list of all lumber and related building supplies so that it can be bid on by several local lumber yards. Select the best lumber supplier before construction begins. The excavator who will dig your basement and your footings will need a two-week advance notice.

The week before construction starts, there are four things you should schedule: excavating, rough grading, digging the footings, and putting in temporary power and water. Water is used primarily by the mason, and electricity is needed by the carpenter. Both trades can bring their own supply, but neither will want to. If there is electricity at your lot, the power company can provide temporary service. You will have either municipal water or a well, and the plumber can provide a temporary tap. Sometimes it is more convenient to make arrangements with a neighbor and pay him for power and water during the early stages of construction.

The second group of items in the third week of the preliminary phase are lead-time items: ordering the structural steel and the rest of the windows, contracting with the mason (or concrete contractor) who will be doing the foundation walls in two weeks, and ordering roof trusses if you will be using them.

Although not on our checklist, this is an appropriate time to consider security. Usually the construction materials on the site are not pilfered during the early stages of construction. Nevertheless, it is a good idea to notify your local police that you have a house under construction and that you will appreciate it if they could send a patrol car by there on a regular basis. If you have neighbors who can see your construction site, you might encourage them to call you or the police if they observe any unusual activity at your house. If you are keeping any tools or small building supplies at the job, lock them in a tool bin or shed. As soon as possible (week 2C) lock the house whenever there are no workmen on the job, and keep it secure for the balance of the construction process.

F. PHASE ONE: THE FOUNDATION

In phase one, our construction schedule checklist is subdivided into three groups within each week or time period. The first group (starting with "house location staked" in 1A) is the physical work that needs to be accomplished during that time period. The second group (starting with "schedule mason" in 1A) is the administrative work you need to get the work accomplished in the next week or time period. The third group (starting with "contract with carpenter" in 1A) consists of the lead-time items that need to be accomplished in order to stay on schedule. These three groups are present in all phases of the schedule.

During the first week (1A) of phase one, the surveyor sets the offset stakes for the house, the site is cleared, and footings are dug and poured. The first inspection is usually the footing inspection, and it is normally done by your county building department that issued your building permit. However, most jurisdictions will permit a registered architect, engineer, or surveyor to inspect the footings. This can be important, because timing is critical. You do not want to open up the ground for the footings and then let it sit for a day or two (or even half a day) waiting for the county inspector. Rain, children, or dogs can collapse the earth sidewalls and force you to redig the footings. You want a minimum amount of time to elapse between digging the footings and pouring the concrete. So if there is any doubt about the county's responsiveness, it may be better to schedule your own surveyor (if the county will accept him), a registered architect, or an engineer to inspect the footings the minute they are dug and certify them to the county.

The first inspection is important, because it insures that the house will have a solid foundation. The footings provide a hard, level surface that spreads the weight of the house over the amount of surface area on the ground. The ground, of course, must have enough strength to hold the house. When you dig the footings, it is important that they be on what is called undisturbed soil, which is virgin earth with no evidence of fill, loose or soft rock strata, or other conditions that appear unstable. If any unstable conditions are in evidence when the footings are dug, consult an architect or engineer to provide an acceptable footing design.

We also recommend the installation of a temporary toilet, for obvious sanitary reasons, unless it is a very remote job site with no reasonably priced services available. And we recommend a temporary telephone, if it is available. This is more for your convenience than your contractors, and it may be well to install the telephone in a locked box to prevent misuse. You will encounter many situations that require immediate telephone communications, and if a telephone is not available in your car it should be there at the job site.

Plan for trash removal. An incredible amount of trash will be accumulated in the construction of a house. A small dumpster placed on the job will be sufficient, although in some suburban areas other people may begin using your dumpster to avoid going to the dump. You can also use large trash cans, barrels, or old oil drums, but you will have to get rid of the trash on a regular basis.

The mason or concrete contractor will need to be contacted to confirm his presence on the job the following week to install the foundation walls.

If you're using a steel girder in the basement, check on the steel delivery ordered in the preliminary phase. You don't have to use steel; you could use wood. But steel will span longer distances than wood, resulting in fewer columns, and fewer columns will mean more usable space in the basement.

If you have not yet selected a carpenter and a plumbing contractor, you'll need to make that decision and sign agreements with them. Groundwork is the pipe that goes underneath the basement slab, and it must be installed and inspected before the slab is poured. It's a good idea to have it done as soon as possible after the first floor deck is complete.

Bathtubs and showers are large and need to be installed early. If you want a special bathtub such as a Jacuzzi, you or your plumber will need to place the order now, because usually it will not be in stock. If you are going to use a metal instead of a masonry fireplace, it needs to be ordered now, to insure availability in four weeks. (The heating efficiency of the two types of fireplaces is about the same, but the masonry fireplace will retain the heat better and continue to warm the house after the fire is out.) A metal fireplace is also considerably less expensive and easier to install. If your house has skylights, order them now.

The second week (1B) is basically a masonry (or concrete) week. Although not on our schedule, we recommend that after the footings are poured and before the mason starts his work, you have the surveyor come back and set points (nails) in the footings at all the corners of the house. We have found him to be more accurate than the mason, and most masons would prefer to have the surveyor set the points.

The mason usually sets the steel, although the carpenters also can do it. We like the mason to install the steel beams, because he is the one who puts in the beam pockets (or holes in the wall) for the steel. In some cases, however, the mason or the concrete contractor will not set the steel, and it must then be made part of the carpenter's contract.

In the second week, you (or your carpenter) should order lumber, and you schedule the plumber.

There is a long lead-time for special hardware. If you saw something in a home magazine that you liked (a nice, darkened-brass front-door handle for instance) you need to check with the company that supplies them, and find out how long it will take to get them. Usually they are not in stock and must be ordered.

In four weeks we hope you will be ready to have the roof installed. It is a good idea to start talking with your roofer now, telling him that you want him there in four weeks with the shingles. Or, if you are ordering them, check to make sure they are available. Usually contractors prefer to buy the material themselves, and they generally will get a better price, which they may pass on to you. Most of the supply houses want to keep the good contractors coming back, so they give them discounts. They don't expect a homeowner to come back. How many houses is he going to be building in a lifetime? The contractors who have a good credit rating with their supply houses are usually the best tradesmen.

Usually the lender will want the wall check survey before he will release the first construction draw. Most first draw schedules end with the first floor deck—the floor joists and the plywood subfloor firmly attached to the foundation walls. This is important to the structure of the foundation and must be completed before any dirt is placed against the foundation walls (backfilling). We have already discussed the plumbing groundwork; if not required by the draw schedule, it can be postponed until the next week.

The last work item in phase one is to ask the lender for your first draw. Usually a phone call is all that is required, and he will have his inspector come out and check the construction. Assuming all the required work is in place, the lender will release the money to you. Sometimes, depending on how your construction contract reads, the check will go to you and your primary contractor. Usually the contractors who have done work up to this point will wait for the draw payment to be paid. However, the excavator and/or the mason will sometimes want to be paid as soon as their work is complete. Payment terms should always be a part of any contractual agreement, and your contractors need to know when they will be paid before they accept the agreement.

If you're ordering the lumber, it is time to order the framing lumber for the exterior walls of the house, the second floor (if you have one), and the roof framing. Waterproofing of the basement walls and installing the sub-drainage material is discussed in phase two and needs to be scheduled now.

We recommend that you contract with an exterminator to treat the soil to keep termites away from your home. In considering the threat of termites, find out what the county recommends. If your building site is in a low and moist wooded area, you will have a greater problem than if it were out in a high, dry field with no trees. Geographically the northern states have a slight hazard of termite infestation, middle states a moderate hazard, and the southeast the heaviest hazard. Alaska is entirely free of termites.

The basement and garage slabs are normally poured after the roof is complete. If you have not reached an agreement with a subcontractor to supply the concrete and to pour and finish your concrete slabs, now is the time to do it.

The term "casework" refers to cabinets and countertops for the kitchen, bathrooms, and any other custom cabinetry to be built into your home. Unless you are purchasing stock cabinets and have confirmed that they are always available, you should allow approximately three months to get the cabinets you want. If you are having a cabinet shop custom-build your casework, it may take longer—especially if the shop is busy.

G. PHASE TWO: THE SHELL

Phase two could also be called "under roof." Most second draws are released after the roof is on, and some will have an intermediate stage after the second floor deck is complete. Our schedule does not allow extra time for the second floor, and will need to be expanded by the time required to frame the house if you are planning an extensive second or even third floor. Technically the first floor deck is part of the shell. If you have one contractor building the shell his work will begin in phase one and may not be finished until phase three.

In the first week of phase two (2A) the exterior walls are framed, waterproofing and sub-drainage is installed and inspected prior to backfill, and the underground plumbing is inspected. Framing the interior non-load-bearing partitions on the first floor is optional at this time. Some carpenters do not want to do that until the house is under roof; others will be willing to do it along with the exterior walls.

The surface of the masonry or concrete foundation needs to be waterproofed (most of the time this is actually dampproofing) to prevent water in the ground from coming into your house. There are several requirements for a dry basement, and waterproofing or dampproofing is just one of them. The difference between waterproofing and dampproofing is in the degree of protection. Both systems start with a smooth surface—either the concrete surface of a poured basement or a coating of mortar called "pargeting" (pronounced parge-ing) which is troweled on over the concrete block. If you are dampproofing (which for most basements is adequate), a coating of asphalt is spray-applied to the foundation. However, if you are faced with a severe subsurface water problem and need to waterproof, a membrane of polyethylene or several layers of asphalt-impregnated roofing felt are imbedded in asphalt to further inhibit water penetration.

In addition to dampproofing, it is important to provide an easy escape route for water to follow away from the foundation. This is accomplished by diverting surface water away from the house (discussed in phase four) and drawing subsurface water below slab level and out to "daylight" by means of subdrainage. At the base of the foundation wall, above the footing and below the level of the concrete slab, a perforated drain pipe is installed in a bed of gravel, with more gravel placed on top and covered with building paper to prevent dirt from clogging the holes in the pipe. This underground gutter will collect any subsurface water that accumulates against the foundation wall. After the basement walls have been surrounded, the drain pipe needs to be discharged to the ground surface at a point lower than the pipe around the footing (daylight). With a walk-out basement this is easily accomplished, but if your basement is totally subterranean, with no lower land close to the foundation, you may have to place the drain pipe inside the house (still below the slab) and discharge it into a sump in the basement floor, with a pump to eject the water outside your house.

You need an inspection prior to backfill. The county will want to see that you have adequate waterproofing and sub-drainage.

You also need to have the underground plumbing inspected before the slab is poured. This inspection is usually requested by your plumber.

If you want a roof ventilator fan, now is the time to order it or make sure that your roofing contractor does. Call your excavating contractor and schedule him for the next week. You or your concrete contractor need to order the material for your basement and garage slabs—the polyethylene and gravel that go underneath the slab, and the reinforcing or temperature steel that goes in the slab to prevent cracking.

The contractor who is placing and finishing the concrete slabs needs to be scheduled for the following week, as does the exterminator.

Lead-time items include ordering the exterior doors and the garage door. If you are having any unusually shaped windows that you cannot order from a window manufacturer, you need to contact a glass company to measure, assemble, and install this glass.

If you have selected a really nice front door and lockset for your house, order it (it will take more than two weeks to be delivered) but do *not* put it in now. Instead, buy the least expensive exterior door and lockset you can find, and install it now. Then, when you are nearing the end of the construction, have your expensive carved wood door delivered, and your carpenter set it for you. As discussed previously, delaying installation will avoid any damage to your expensive door during construction, while at the same time giving you the security of a door you can lock.

If you are going to be using a brick or stone veneer on the exterior of your house, select and order it for delivery in three weeks.

During the second week of phase two (2B), the carpenter should finish framing the house and the roof. If you're building a large two- or three-story house, the time required to complete the framing will probably exceed the one week provided in the schedule. A large, complicated contemporary home can sometimes take four to six weeks to frame and sheathe.

After you have installed the sub-drainage and had it inspected (if required), the backfill can be placed carefully and in compacted layers into the excavation. All the basement and garage slab preparation work can now be accomplished, and the slab poured. Again, an inspection may be required to assure the county that the slab will be structurally sound. This is particularly critical if any portion of the slab is being constructed on fill dirt.

Schedule your roofing contractor for next week, and check the availability of the shingles or other roofing material. If you are going to put framing in your basement, you will need to order more lumber. All exterior doors should be available next week, and the garage door contractor should be scheduled.

Lead-time items include completing your agreement with the mechanical contractor for your heating and air-conditioning system. If you find some really attractive electrical fixtures—a chandelier, a front entrance light, or a lantern for your front walk—your local supplier will probably need two or three months to order it for you. All recessed lighting fixtures and exhaust fans have housings that are installed in the framing at rough-in, and should be ordered now. Your electrical contractor may be ordering them for you, but make sure they will be available in four weeks.

In the third week of phase two (2C), your roof will be completed, and any basement partitions can be framed, all exterior doors can be installed, and the house can be locked. This will provide some security during the balance of the construction period.

If this schedule corresponds to your draw schedule, you can now request that your lender inspect the work and release the next portion of your construction loan.

For next week's work, you need to schedule the plumbing contractor and order the material for the exterior siding and any fixed-glass frames. If the exterior has a brick or stone veneer, you need to schedule the mason.

H. PHASE THREE: THE MECHANICALS

Progress during phase two will be very obvious to the observer, whether a neighbor or someone who drives by every day, because the shell goes up very fast. During the next phase, however, most of the work goes on inside the house, so that sometimes it will seem that nothing is happening during phase three, which can take six weeks or longer. In the first half of phase three, you will be working mostly with the mechanical trades—plumbing, heating and air conditioning, and electrical.

As always, your biggest problem will be getting the tradespeople on the job to work in the proper sequence. Rarely will they be able to be on your job the day that you want them there, but they will show up the day following, or the day following that, or. . . . However, your diligence at keeping after them will pay off as the rough, empty interior shell becomes a house, with rooms, doors, and cabinetry.

The plumbing work is the least flexible of the mechanical trades as to location and size, and therefore must come in the first week of phase three. At the completion of his work the plumber will call for an inspection. Meanwhile, carpentry work can continue on the exterior of the house, with the framing of any fixed-glass openings and the installation of exterior siding. The glass company will need to be scheduled for the next week to install the fixed glass, and will probably want to measure the opening(s) during the week in order to have it ready for next week. The HVAC (heating, ventilating, air conditioning) contractor needs to be scheduled for the following week. If the masonry walls in the basement are to be finished you need to order furring for the carpenter to install. Complete your agreement with the electrical contractor if you have not already done so, and notify him of his starting date in two weeks.

The second week of phase three (3B) belongs primarily to the HVAC contractor, with the carpenter or mason doing some exterior work and/or furring out the basement walls. Note that the HVAC work is not inspected until after the insulation is complete, since the latter is equally important to the successful heating or cooling of your house. Check with your electrician to confirm that he will be on your job next week.

After the plumber, HVAC contractor, and the electrician have finished their rough-in work, there will be some framing members that have been cut or eliminated, and there will be plumbing, duct work, and wiring that may have to be boxed in. The extra framing work necessitated by these conditions is referred to as back-up. Cabinetry, recessed bath accessories, and some partition framing will require blocking. This material will need to be ordered by you or your carpenter for installation after the rough-ins are complete.

By the end of the second week of phase three you should have agreements with both your insulation and drywall contractors, who will be working for you during the fourth, fifth, and sixth weeks of this phase.

The third week of phase three (3C) starts with the electrician doing his rough-in work, including the installation of the recessed light-fixture housings we mentioned earlier, and getting his work inspected. As soon as the electrician is done, the carpenter will need to install the blocking and back-up work and make sure everything is ready for the framing inspection. Depending on how exacting your inspector is in enforcing the county's building code or his interpretation of it, you may pass the framing inspection the first time or be given a list of corrections he wants before accepting the framing. This is usually the most difficult inspection to get through and rightly so; the city or county is doing its best to protect you and any future occupants from unsafe construction.

If your carpenter or mason has not finished the exterior siding or veneer, he still has this week in which to do it. Schedule the insulation contractor for the following week, with the expectation of passing the framing inspection. You or your carpenter will need to order the exterior trim material—the fascia and soffit, and any decorative window and door trim, cornice and/or corner trim.

Concluding your work in the third week is the ordering of your finished stairs and the fixtures and fittings for the baths, powder rooms, and kitchen, especially if there are any special colors or unique items that are not kept in stock. Using the same principle we applied to the front door, have your carpenter build temporary rough stairs during Phase Two, and then replace them with the permanent stairs during Phase Four. There will be a lot of construction traffic between all levels of the house, and you do not want your good hardwood stairs to be damaged. Place the order and make arrangements with the stair company to measure for the stairs before they build them.

I. INSULATION AND DRYWALL

The fourth week of phase three (3D) begins with the installation of the insulation, followed by the HVAC inspection. The next item on your checklist deserves special attention. We have included a concealment checklist in four parts that must be completed before going any further with the construction of your house. At this point, you become the primary inspector of your house, and you should require your four principal contractors to go through their part of this checklist with you. This is the last chance you will have to make sure that all the work that will be covered by drywall and/or other finishes is installed correctly and, where necessary, identified for future reference. If you learn nothing else from this book except to use this checklist, you will have gotten your money's worth. Many people contributed to this checklist, and time and time again it has proved to be of inestimable value. Some of the items may not apply, but it is extremely important that all of them be considered, and if applicable, done.

The framing carpenter, the plumber, the HVAC contractor, and the electrician each has a section. If you are not sure what an item on the checklist means, ask them. Go through it with them before you proceed any further.

CONCEALMENT CHECKLIST

Framing
- ☐ All walls and openings per dimensions on plans.
- ☐ All studs plumb within tolerance—especially at cabinets, vanities and countertops.
- ☐ Kitchen and bath soffits (bulkheads) stable, level, of proper dimension and with adequate nailing.
- ☐ Nailing provided for wall-mounted cabinets, wall-hung fixtures and for any special conditions.
- ☐ Nailing provided for drywall ceiling installation.
- ☐ Nailing provided for drywall corners and stop beads.
- ☐ Nailing provided for all special or unusual conditions.
- ☐ Tub access framed to receive panel and finish trim.
- ☐ Attic access framed to receive panel or stairs and finish trim.
- ☐ Mirror cabinet roughed-in to correct size, location, and not in conflict with light or switches.
- ☐ Recessed bath accessories roughed-in or location clear of interference.
- ☐ Door rough openings correct for size and type of head/jamb—extra height provided at carpeted areas to avoid undercutting.
- ☐ Frame provided for electrical panel if in finished area.
- ☐ Changes to framing for rough-ins properly corrected—sound, plumb and level.
- ☐ Window-jamb extensions to proper dimension to be flush with drywall.
- ☐ Window head, jamb, and sill plumb or level to receive proper trim thickness.
- ☐ Vertical chases properly firestopped.

Plumbing
- ☐ Under-slab stub-ups in correct locations
- ☐ Floor drain roughed-in at ☐ furnace, ☐ HWH, and ☐ laundry.
- ☐ Ice-maker supply line roughed-in with coil.
- ☐ Exterior hose-bibs and interior shut-off valves in accessible location.
- ☐ Sufficient space for ☐ HWH, ☐ clothes washer, and ☐ clothes dryer.

☐ Accessible location for meter yoke.
☐ Remote water meter roughed-in.
☐ Water closet tank supply line 8″ above finish floor line.
☐ Tubs protected.

Mechanical
☐ Bath exhaust fan roughed-in with duct to outside.
☐ Range exhaust duct in proper place.
☐ Clothes dryer exhaust duct in proper place.
☐ Supply and return openings installed per plan or located within acceptable practice.
☐ Ductwork brought to proper location for air handler.
☐ Ductwork properly insulated.
☐ Thermostats roughed-in.

Electrical
☐ Wiring to ☐ range, ☐ disposal, ☐ HWH, ☐ all air handlers, and ☐ all condensors or heat pumps, ☐ garage door opener.
☐ Lighting junction boxes roughed-in per plan, properly centered, level or on axis.
☐ Boxes do not project beyond studs, joists, or rafters more than finish thickness, preferably less.
☐ Boxes installed plumb or level.
☐ Multiple switches in correct order for convenient operation per plan.
☐ Switches at latch side of door and not in conflict with trim or cabinetry.
☐ Telephone wiring and rings in proper place per plan.
☐ Kitchen wall phone not in conflict with wall cabinets or window trim.
☐ TV cable and/or antenna wiring and rings in proper place per plan.

Returning to the construction schedule checklist, we see that during the installation of the insulation inside the house, the carpenter can be working on the exterior trim. You need to schedule your drywall contractor and make sure the drywall material has been ordered. If you are building in winter, you or your drywall contractor must arrange for temporary heating, and make sure the house is reasonably airtight. The spackling compound used with the drywall needs to be kept at a minimum of fifty-five degrees for proper drying.

During this week, you also need to contract for ceramic tile (if applicable), and order the balance of the finish hardware, including the interior door latches and closet door pulls.

In the fifth week of phase three (3E), while the drywall contractor is installing his work, you need to order electrical fixtures and hardwood flooring for future deliveries, and arrange for the drywall finishing if you have split the drywall work into two contracts. (Although you may save a little money with two separate contracts, unless you know a good finisher it is probably better to have a reputable drywall contractor do the entire job.)

The same principle is true for the hardwood and other floor finishes. You can save money by contracting separately with each tradesman, but your management time increases significantly. There are good flooring contractors who will do hardwood, ceramic, vinyl, and carpet. You will still need to select your finishes and schedule the installation of each product in accordance with the checklist. At this time it is important to select the carpet (and the carpeting contractor if applicable) and make sure it is ordered so that it will be available when needed.

In the sixth and final week of phase three (3F), the drywall is taped, spackled, and sanded. The exterior trim should be completed this week if it hasn't been done in previous weeks. And, if appropriate, request your next draw from your lender.

The first flooring material to go in your home is unfinished hardwood, and should be scheduled for installation next week. Interior trim lumber (doors, base molding, window trim, and other required items need to be ordered, and the finish stairs should be completed and ready for delivery.

Schedule your HVAC contractor to come back to set the grilles and registers, and your electrician to install outlets and switches.

Three contracts need to be completed during this week. Gutters and downspouts are usually done by a separate contractor, and should be installed in about two weeks. Paving may be stone, asphalt, or concrete, and will need to be installed in about three weeks. If your resilient flooring (vinyl sheet or vinyl tile) is being done by a separate contractor, you will need an agreement for installation in about three weeks.

J. PHASE FOUR: INTERIOR AND EXTERIOR FINISHES

The first week of phase four (4A) begins with hardwood flooring. The installation of the unfinished floors precedes the installation of the doors and interior trim, because the doors and base are set on top of the hardwood. If you are using prefinished hardwood strip or parquet flooring, that should be installed but carefully protected. The other floor finishes (vinyl and carpeting) are fitted to the door frames and base trim and installed later.

In addition to hanging the interior doors and installing the trim, the finish carpenter will also install your finished stairs. The stairs should be protected during the final construction phases. The HVAC contractor will install the grilles and registers, and the electrician will install the outlets and switches. Although most tradesmen will prefer to work separately, the HVAC contractor and the electrician can work together at this point in the construction without being in each other's way. Many of the remaining weeks will have more than one trade working simultaneously.

Schedule the painting and order the paint if you are supplying it. Also schedule the installation of gutters and downspouts. If your house has exterior wood decks, the material needs to be ordered. Also, order material for stair and interior balcony railings and handrails, and underlayment for all vinyl and ceramic (master set) floors. Check with your hardware supplier to make sure all of your hardware is available to be installed next week.

If you are having a separate contractor sand and finish the hardwood flooring, he needs to be under contract and ready to start in two weeks. Bath accessories will be needed in about four weeks.

When the interior trim is finished—ideally near the beginning of the second week of phase four (4B)—the interior painting can begin. No other interior work can be done during the painting, but outside the house the decks, stairs and railings can be constructed, and any exterior trim not already in place can be completed. When that is done, the painting and staining work can move from inside the house to outside.

Then, inside the house, the stair and balcony railings and handrails, the finish hardware, and the underlayment can be installed by the carpenters.

After the exterior painting is complete, have your gutters and downspouts installed. Many counties will permit you to eliminate the gutters if your roof overhang is at least two feet. However, water running off the roof will soon create a gully in the ground under your roof overhang, and mud will splash against your house. One way to control this is to dig a trench on the ground under the roof and fill it with gravel or rocks. Even then, however, some water and dirt will splash on the side of your house, and there is no guarantee that the water will not find its way back to the foundation and into the basement. So gutters and downspouts, properly installed and with splashblocks (or preferably undergound piping that leads the water at least five feet away from the house) are the best way to maintain clean siding, and one of the important factors in achieving a dry basement.

If you need retaining walls around your house, walks and/or driveway, and you are going to have them constructed with railroad ties or a similar treated wood product, order that material this week, or contract with a landscaping contractor or your grading contractor to do the work.

The scheduling this week for work to be done next week includes the contractor who will pave your driveway, the tradesman who will install the ceramic tile, the tradesman who will install your resilient flooring, and another who will finish your hardwood floor. You should also schedule the delivery of the vanities and kitchen cabinetry (referred to in the schedule as casework) that you ordered eleven weeks ago.

If you have purchased a special wood front door and want it properly weatherstripped, contract with a company that specializes in installing spring bronze weatherstripping and a compressible-gasket threshold. If you have purchased pre-hung doors you may not need this service. After your casework is installed you will need countertops, and these should be ordered for delivery in two weeks. The company supplying plastic laminated countertops will want to take exact measurements *before* fabrication. They may want to wait until the cabinets are installed, so schedule the measuring for next week. Finally, order your kitchen and laundry appliances for delivery in four weeks. Most good distributors should be able to meet that deadline.

In the third week of phase four (4C), the exterior work includes finishing the grading and installing the splashblocks of drain pipe for the downspouts and paving the driveway. Interior work includes the installation of ceramic tile floor and tub surround (if applicable) in your bathrooms, installation of the resilient flooring, finishing the hardwood floors, and installing the cabinetry in that order. At the same time you need to schedule the concrete finisher for your exterior flatwork—patios, walks, and possibly the driveway, the installation of weatherstripping and thresholds, the delivery of your kitchen countertops, and the installation of the electrical cover plates and the internal wiring and circuit breakers in the panel box.

The minimum landscaping requirement in most counties is for some form of ground cover over all areas of exposed dirt. A growing ground cover such as grass (as opposed to bark mulch) is highly recommended around the house in order to retain ground water and keep the foundation walls dry. Foundation planting is also desirable for the same reason. Keeping the grade around the house sloping away from the house and having living roots in the soil are also very important factors in achieving a dry basement. We hope that your budget will allow for some nice plantings on your property in addition to the minimum required ground cover. Contract for this work to be installed in two weeks.

By this time, you should certainly have a feeling for the construction of a house. Each week, while the tradesmen are doing their job, you have been looking over their shoulder, while at the same time ordering material, contracting, and scheduling.

By the fourth week of phase four (4D), your house looks as if it were nearing completion—as indeed it is! During this week the patios and walks are finished and the weatherstripping, thresholds, kitchen counter tops, electrical cover plates, and panel internals are installed. After this work is complete, and if appropriate, request your next draw from your lender. Scheduling includes the seeding, sod and/or mulch, the installation of sinks, water closets and fittings, the exterior heating and air-conditioning equipment, and the electrical fixtures. Check to be sure that the bathroom accessories and fixtures are available for delivery next week.

K. PHASE FIVE: COMPLETION

This is the beginning of the final month of construction and the end is in sight. Contact your surveyor and have him prepare the final plat, which will include everything you have added since the wall check survey—the driveway, the walks, patios, the decks and the retaining walls. The final plat is usually required by your lender for final payment or by the county for occupancy, but if neither requires one you can skip this item. Exterior work includes seeding, sodding, and planting, and installing and wiring the HVAC condensing unit and any outside light fixtures. Interior work includes installing the bathroom sinks, water closets and fittings, and the electrical fixtures. During this first week of phase five (5A), to schedule the delivery of your appliances, schedule the final hook-up testing, and balancing of the HVAC system, and request the final plumbing, electrical, and HVAC inspections.

Throughout this chapter, we have discussed the various construction techniques that contribute to keeping your basement dry. They fall into two categories: providing an easy escape route for water to follow away from the foundation both on the surface of the ground and below it, and protecting the foundation walls and slab from any remaining water. Only when all of these components are in place (at the end of this week of construction) can you expect your basement to be dry.

The last request you make of the painting contractor is for him to return in two weeks to do interior and exterior touch-up, and to paint the gutters and downspouts if necessary. You also may want to consider hiring a cleaning service to do the final clean-up, which will be a formidable task. Cleaning a newly constructed house is very different from cleaning the house that you live in all the time. There are companies that specialize in this work. Not only do they vacuum, dust, and mop, but they also remove all the dirt, paint splatters, labels, and stickers from the countertops, the sinks, the water closet, the tub, the glass, the vinyl flooring, and the ceramic tile.

In the second week of phase five (5B), the major appliances will be installed. Usually the county has a sequence for the final inspections, starting with the final plumbing inspection. Before the final HVAC inspection, have your contractor balance the system—that is, make sure that the airflow is adequate to all registers and that the dampers in the ductwork are adjusted properly (assuming you have a forced-air system). Walk around the house with your contractor and check the amount of air coming out of the registers, and make sure that all rooms are getting their share of conditioned air (check especially any remove rooms or rooms over a non-heated space like the garage). After the final HVAC inspection, the next final inspection is electrical.

Schedule the touch-up painting and the painting of the gutters and downspouts, schedule the cleaning contractor, and schedule the carpet installation.

In the third week of phase five (5C), after the paint touch-up work has been done, the entire house cleaned, and the carpeting installed, and after *you* have checked all work for completion, schedule the final building inspection and request the final plat from your surveyor.

In the fourth and final week of phase five (5D)—ideally, but most unlikely, twenty weeks after you began construction you have a house that you can live in. After the final building and public utilities inspection, your county should issue your occupancy certificate. And, of course, you can request your last draw from your lender.

L. THE FIRST YEAR

By way of concluding this chapter, we want to alert you to some things that may occur during the first year you occupy your newly constructed home. All the material that you have incorporated into the construction of your home has come to the site with a high moisture content. This is particularly true of the lumber products because of their molecular structure. But it is also true of inorganic material like drywall. During the first year these materials lose most of their "excess" moisture, especially during the heating season when the relative humidity inside your house can get down into single-digit numbers.

As these building products lose moisture most of them will shrink. Sometimes exposed wood products like beams, columns, wood decking, and even trim will twist or develop small cracks along the grain called "checks." These are not structural failures but characteristic action of wood as it drys. Concrete slabs and concrete masonry may also develop small hairline cracks as the material dries. Again, these are not structural failures. Drywall will usually show "nail pops" within the first year (due primarily to the wood studs drying and shrinking) and these simply require spackling and touch-up painting. If the drywall seems loose and not adequately fastened to the wood studs you should add a nail or screw as required, spackle, and touch up.

All of the above are normal occurrences and should not be a cause for concern. However, if you find large or long cracks in concrete masonry walls or any finish material and/or doors or windows that bind in their openings when they had originally worked properly, these may indicate a structural problem or a differential settlement in the foundation. It is probably best to have a professional engineer, architect, or home inspector check these conditions. Usually they will stabilize after the first year and will need only cosmetic repair.

10 Mechanic's Lien Law

This definition comes from Black's *Law Dictionary*:

> A Mechanic's Lien is a claim created by law for the purpose of securing priority of the payment of the price of value of work performed and of materials furnished in erecting or repairing a building or other structure and as such attaches to the building as well as to the land.

In other words, a mechanic's lien is a claim against real property created by having performed work or services or having provided material there. It's a priority claim, and this is very important. The law says that someone has given value to your home, and that it is unfair for that person to come after someone else who might claim your home as an asset, since he's the one that created the asset for you. The actual physical builder—the one who lays the bricks, frames the roof, installs the flooring—comes first. Without him there would be no home to borrow money on. That's why the mechanics have a higher priority than most other claims.

A. THE PLAYERS

The mechanic's lien laws vary from state to state, but the players are usually the same, and it is important to know them. They are:
  (1) The owner
  (2) The general contractor
  (3) The subcontractors, including:
      (a) Mechanics: plumbers, electricians, drywall men, carpenters, etc.

(b) Materialmen: the companies and individuals who supply material such as drywall, lumber, cinderblocks, etc. (Under the law, there is no distinction between mechanics and materialmen.)

(4) The lender

(5) The Internal Revenue Service (to some extent).

B. WHO CAN FILE A LIEN?

The general contractor, the materialmen, and a mechanic who is a subcontractor can file liens on your property. Judgment creditors and the IRS can also file a lien on your house, but they will not have the potential priority that a mechanic has.

What about an employee of a contractor or subcontractor who has worked on your house and is due money from the contractors or subcontractors? Can he file a lien on your house? No. Employees cannot file a lien on a house. Contractors and subcontractors can, but not their employees.

If you are an employee of a contractor, your claim for nonpayment is against the contractor. This is important, because there will be times when there is a dispute between the contractor and his employees, and frequently an employee will call you and say: "I'm John. I worked on your house and I haven't been paid in a month. So I'm going to file a lien against your property." If so, you need not worry, because employees have no lien rights.

C. TYPES OF LIENS

There are two types of liens: voluntary and involuntary. A voluntary lien is created when you borrow money to build or to buy your house. You give the lender a voluntary lien on your property.

There are three types of involuntary liens.

(1) If you don't pay your taxes to any jurisdiction, you will have a lien very quickly. It could come from the federal government, the state government or the city government.

(2) If you are sued for something concerning your business or personal affairs, and somebody obtains a judgment against you, that judgment has nothing to do with your house. But if you do not pay up, within a very short period of time that judgment can be filed in the county where you live, and if it is a personal judgment against you the judgment becomes attached to your house.

(3) Another involuntary lien arises when a mechanic or materialman who performs work or provides services or material is not paid and files a lien.

D. LIEN PRIORITIES

Generally speaking, the lien with the highest priority (the highest claim to the property) is the first deed of trust (or the mortgage). It follows that the second lien filed against your property is called the second trust and right on down the line. It also follows that the higher your lien priority the greater likelihood that you will be paid the money due you if your property has to be sold.

Remember, a lien may be voluntary or involuntary. Accordingly, if the IRS filed a lien against your property, their lien priority would be based on their place in line.

Here's how it works. You borrow $100,000 to build your home. The lender files his lien, which is the first trust. Then you run short of money and borrow $10,000 from a friend, who now has a second trust. You forget to pay your taxes, and the IRS files a lien on your property. It has a third trust. You don't pay the plumber and he files a lien. His lien is in fourth position. It is easy to determine lien priorities: first come, first served.

There is, however, one very important exception applicable in some states, although it is not as important to you as it is to the lender. This exception is called "piercing the trust." If a mechanic files a lien and it ultimately proves to be valid, that lien will jump ahead of the construction loan first deed of trust. It's one of the few exceptions to the priority rule.

The IRS must stand in line like everybody else, but in a little more than half the states in the country a mechanic can jump ahead in front of the line. As a result, when a lender in some states advances money to you on your construction loan, he runs the risk that his money will not be in a first trust position. If somebody does $10,000 worth of work on your property and you don't pay him, and he files a lien and ultimately sues and wins, he can foreclose on your property. The first $10,000 from the sale of your property will go to the lien holder. In states that do permit piercing the trust, the lender is now in second position. This is done to protect the contractors. In states where piercing the trust is not allowed, when a contractor files a lien it is subordinate to the total amount of the loan. So if you are a lender in a state that permits piercing the trust, you could be put in a subordinate position every time a contractor files a lien. Thus, in these states, every time a lender advances money he requires that the lawyer who does the title work go to the courthouse and check the title to see that a lien has not been filed. This is called a title bringdown.

Only when the lawyer notifies the lender in writing that there are no liens will the lender advance the money. Usually the lender will also require that you secure a lien waiver from the contractor, documenting the fact that you have paid him and that he has paid his subcontractors. Requiring a lien waiver and a title bringdown is double coverage. If you do not have a lender, you should go through the same process to protect yourself.

In piercing-the-trust states, lenders usually require affirmative mechanic's lien coverage (AMLC). It protects the lender's position in the event a valid lien is filed and attached to your property.

Only the lender is protected by an affirmative mechanic lien coverage. There is no insurance that will protect you from somebody filing a lien against you. You must protect yourself.

E. HOW THE SYSTEM WORKS

Here is how a mechanic's lien is usually filed. The creditor, who could be the general contractor, a materialman, or a mechanic, has a certain number of days (generally 90) after the performance of the work or the completion of the home (whichever occurs later) in which to file a lien. He fills out a document that is very technical and records it at the courthouse.

The creditor also files a lawsuit. He has a certain period of time after he files his lien to file the suit, and here is where most liens die. In the heat of an argument the contractor goes down to the courthouse and files a lien. But after thinking it over he says: "Hell. I'm not going to pay a lawyer $100 or $200 an hour to sue some guy for $1200." So even if you owe him money, he will very likely never get around to filing the lawsuit. And after the prescribed time for filing has lapsed, if he has not filed the lawsuit, it lapses by law. It will not necessarily eliminate the claim against you, but does extinguish his lien on your property.

This is how it works in most states. The exception is in states where the lien and lawsuit could pierce the trust, threatening the lender's first position. In those states the lender sometimes will not advance any more money until the matter is resolved. Resolution for the lender could come either by settlement of the matter between borrower and contractor or the borrower securing a bond protecting the lender if the lien is held valid. With this bond the lender is assured and can continue to advance funds to the borrower from the loan. Check you state law to see if a mechanic's lien can pierce a trust. Then, check the time frames in your state for filing a lien and for filing suit to perfect the lien.

F. LIEN WAIVERS

1. Contractors

Since liens are potentially foreboding, the system has developed an orderly process for dealing with liens and potential liens. It is called a lien waiver.

When a mechanic performs work on your home, he has the right to file a lien until he has been paid. When he is paid he then waives this right.

Each time you pay a mechanic you should have him sign a lien waiver for the work he has done and for the amount he has been paid. This lien waiver is added protection for you, as well as assurance for your lender that you have paid the mechanic.

2. Subcontractors

Suppose you pay the general contractor, but the general doesn't pay the subcontractor. This could cause problems. To be super careful, you should require that your contractor provide you with a lien waiver from the subcontractor to extinguish any lien rights on your property.

G. A TALE OF THREE PLAYERS

Even if you pay all of your bills, if you are not careful, you could still end up with a valid lien filed against you. The potential problem comes from the owner/contractor/subcontractor relationship. Keep in mind that anyone who contracts with the owner, regardless of what he does, is a general contractor. In the same way, anyone who contracts with the contractor to do work for the owner is a subcontractor.

Suppose you contract with a general contractor to build your house. He then hires a plumber to install the plumbing. The plumber plumbs your house. The general contractor submits a $2,500 bill to you for the plumbing. You inspect the plumbing and pay him $2,500. The general contractor gives you his lien waiver for $2,500 for the plumbing. What happens if the general contractor does not pay the plumber and the plumber files a valid lien?

You are not liable unless you had notice or knowledge that the plumber had not been paid by the general contractor when you paid him. If you did have knowledge, either actual or constructive, that the plumber had not been paid, and you paid the general contractor despite this knowledge, then the plumber's lien would prevail. If the plumber's lien prevailed and the general contractor were able to get by without making the payment good, you would end up having to pay twice for the plumbing.

The key is not to pay the general contractor when you have actual or constructive notice that he owes money to a subcontractor. What is actual notice? When the plumber, whom you have never met, calls you at night and says, "I haven't been paid for plumbing your house," that's actual notice. Do not pay when such actual notice exists.

What is constructive notice? When the plumber files a lien on your property, that's constructive notice. So before you pay a general contractor, check the courthouse to see that no liens have been filed. If no liens have been filed and you have no reason to think that the plumber has not been paid, then you can pay the general contractor with some confidence. In a trust-piercing state, the lender will go through the same process to protect his first trust state.

There is another point to remember. Anyone can file a lien. But if the contractor files a lien for improper reasons, it is slander of title, and he could be sued for hundreds of thousands of dollars. If you are going to file a lien, the law says you must be positive you have a valid claim. You do not have to win, but you cannot go around arbitrarily filing liens and suing people merely as a way of threatening them.

H. HOW TO PROTECT YOURSELF

First of all, always pay by check. Identify what you are paying for on the check. When you pay, ask for a lien waiver, a waiver that surrenders the right to file a lien based on the amount of the check you are giving the recipient. If you pay the contractor $2500 and he has done $2500 worth of work, and you have a check, it is evidence of payment. So it is highly unlikely that he will file a lien. Every time you make a payment, ask the contractor to sign a lien waiver for the amount of the work that he has done since the last payment, and require lien waivers from the subcontractors if there are any. Your lawyer can provide you with all the lien waiver forms you need.

There are two types of lien waivers: the short form and the long form. There are legal differences between these two that are not important to you. The main practical difference is that the long form has a place for all the contractors and subcontractors to sign while the short form is usually for the contractor alone. With the long form you can end up with one form and fifty names; with the short form, you might end up with fifty forms. The problem with the long form is that you might have forty-five signatures and then lose the document. With each additional name on the long form the document becomes more critical. The loss of one short form waiver is only that: one missing waiver. Hence we recommend dealing with separate lien waivers for each contractor.

I. A FOOTNOTE

Suppose you owe your general contractor $2500 for plumbing. Suppose you hear that the plumber is owed $2000 by the general contractor. The general contractor asks for his money, and you say that you will not pay him until he provides proof (lien waiver) that the plumber (subcontractor) has been paid. He replies he cannot pay the plumber until you pay him. This is not an infrequent experience. The general contractor also says that he cannot continue without payment.

Is your project now dead in the water? Not at all. Write a check for $2000 payable to both the general contractor and plumber and write another for $500 payable to the general contractor alone. Ask for lien waivers from both. Now everybody is paid, you have your lien waivers, and the project can proceed. Caution: Do not write a $2000 check made out only to the plumber. There may be a genuine dispute between the general contractor and the plumber. Your contract is with the general contractor. If you pay only the plumber, you could lose. So pay them both with one check, and let them resolve it.

11 Legal Aspects of the Loan

A. THE DEED

There are three kinds of deeds:

1. The Quitclaim Deed

In a quitclaim deed the seller deeds to you any and all interest that he may have in the property. For example, you could give a quitclaim deed to your neighbor for the Holiday Inn you stayed in recently, without having done anything illegal. You could have conceivably even charged money for the deed. All you were doing was deeding any and all interest you have in a Holiday Inn, which was zero interest. That is a quitclaim deed. It says only that you quit or relinquish any future claim to something. You'll see that from time to time when there is a suspicion that somebody may have a claim to the title.

2. A Special Warranty Deed

Under this deed, the grantor or seller certifies to you that he has personally done nothing to cloud the title on a property you are buying. But he doesn't certify that he took good title. There may have been prior title problems. State regulations on deeds vary, but a special warranty deed is not as comforting to the buyer as a general warranty deed. If you're taking title to property by special warranty deed, obtain assurances from your lawyer or your title company that your property is free and clear of all title problems.

3. The General Warranty Deed

The deed which offers you the most protection is the general warranty deed. In this deed the grantor (seller) warrants that not only has he done nothing to cloud or damage title to the property, he warrants that he is conveying good title. Unless your state is an exception, the deed through which you obtain title to your property should be a general warranty deed.

B. THE NOTE

When you go to settlement on a loan, you have to sign a note. A note is a promise to pay a certain sum within a certain time frame. The note we have included in this book is a typical FNMA (Federal National Mortgage Association) note. (See pages 200–207.) If you look at it, you'll see that it has spaces for the amount, the bar, where to pay, the name of the payee, and the lender. It gives the amount of the loan, the interest rate, and tells where to make the payment. This information is included because the lender will probably sell the note to the secondary market.

Look at #4. The borrower may prepay the principal amount outstanding, in whole or in part. This is a guarantee that there is no prepayment penalty associated with this note.

Note the paragraph which says that the borrower shall pay the noteholder a late charge (#6). That is sometimes a negotiable item.

C. THE DEED OF TRUST (MORTGAGE)

Now look at the last line of the note. The indebtedness evidenced by this note is secured by a deed of trust. This date will be the same day as the note. If you borrow $100,000 from a lender and sign a note, and then do not pay it and move to Mexico, the lender is out of luck—except for one clever thing that the lender does. When he loans you the money he also takes an interest in your property (a voluntary lien). You own the property by deed, and he puts a lien on your property with a deed of trust (mortgage). What you're doing is deeding your interest in the property to the trustees. There will probably be two trustees here, who may or may not be lawyers. They will hold this property in trust until such time as the note is paid off. When the note has been paid in full, the trustees will release the lien.

We have included a sample Deed of Trust. Notice paragraph two. Some lenders require that taxes and insurance be put in escrow, while some do not. This is usually a negotiable item. But deeds of trust always require insurance because the lender insists on being protected. If you don't buy it, he's going to do so and charge you a premium.

Look at Paragraph 6. You are not permitted to let your property go to waste. If you do, the lender can accelerate the note—that is, force you to pay it off.

Now note paragraph 17, assumability. If a contract, a note or a deed of trust does not specifically prohibit assumability, then it is permissible. But always make sure it's assumable if you plan to sell your house before paying off the mortgage.

D. LEGAL VS. EQUITABLE TITLE

Legal title is confirmed by the deed. The grantee on the deed is the legal owner of the property. When you execute a contract to purchase a parcel of property from the legal owner and when the legal owner accepts your offer, you become the equitable owner of the property. The distinction between legal title and equitable title is one of degree. The important thing for both buyer and seller to remember is that once a contract of sale is finalized there are now two parties who have an interest in the property. The legal owner (seller) and the equitable owner (buyer). After you've gone to settlement and the transfer of the land has been recorded at the courthouse, you then have legal title to the property.

The trustees in the deed of trust do not have legal title to your property, but they do have equitable title. The difference between equitable title and legal title is one of degree. People with equitable title, however, have some interest in the property.

E. THE MECHANICS OF FORECLOSURE

When you execute the note to purchase your property (or home) you will execute a deed of trust or a mortgage, which as we have seen, secures the note by assuming an interest (equitable title) in your property.

What happens if you stop paying on the note?

At that point the lender may, at his option, accelerate the note. This means that the full amount of the note is due. But the lender must give notice. He must give the borrower a certain amount of time to pay the note in full.

If you don't pay off your note in full by the prescribed date, he can then foreclose on the property. The lender will advertise that the property is for sale. The property is then sold and the proceeds used to pay off the lienors. First lien, second lien, and so forth, in order of precedence.

[Space Above This Line For Recording Data]

DEED OF TRUST

THIS DEED OF TRUST ("Security Instrument") is made on ....,
19 .... The grantor is .... ("Borrower"). The trustee is ....
of ...., Virginia, and .... of
Virginia, trustees (any one of whom may act and who are referred to as "Trustee"). The beneficiary is ...., which is organized and existing
under the laws of ...., and whose address is .... ("Lender").
Borrower owes Lender the principal sum of ....
.... Dollars (U.S. $ ....). This debt is evidenced by Borrower's note
dated the same date as this Security Instrument ("Note"), which provides for monthly payments, with the full debt, if not paid
earlier, due and payable on ....
This Security Instrument secures to Lender: (a) the repayment of the debt evidenced by the Note, with interest, and all renewals,
extensions and modifications; (b) the payment of all other sums, with interest, advanced under paragraph 7 to protect the security
of this Security Instrument; and (c) the performance of Borrower's covenants and agreements under this Security Instrument and
the Note. For this purpose, Borrower irrevocably grants and conveys to Trustee, in trust, with power of sale, the following
described property located in ...., Virginia:

which has the address of .... , ...,
[Street]

Virginia .... ("Property Address");
[Zip Code]   [City]

TOGETHER WITH all the improvements now or hereafter erected on the property, and all easements, rights, appurtenances, rents, royalties, mineral, oil and gas rights and profits, water rights and stock and all fixtures now or hereafter a part of the property. All replacements and additions shall also be covered by this Security Instrument. All of the foregoing is referred to in this Security Instrument as the "Property."

BORROWER COVENANTS that Borrower is lawfully seised of the estate hereby conveyed and has the right to grant and convey the Property and that the Property is unencumbered, except for encumbrances of record. Borrower warrants and will defend generally the title to the Property against all claims and demands, subject to any encumbrances of record.

THIS SECURITY INSTRUMENT combines uniform convenants for national use and non-uniform covenants with limited variations by jurisdiction to constitute a uniform security instrument covering real property.

VIRGINIA—Single Family—**Fannie Mae/Freddie Mac UNIFORM INSTRUMENT**   Form 3047 6/86

UNIFORM COVENANTS. Borrower and Lender covenant and agree as follows:

1. Payment of Principal and Interest; Prepayment and Late Charges. Borrower shall promptly pay when due the principal of and interest on the debt evidenced by the Note and any prepayment and late charges due under the Note.

2. Funds for Taxes and Insurance. Subject to applicable law or to a written waiver by Lender, Borrower shall pay to Lender on the day monthly payments are due under the Note, until the Note is paid in full, a sum ("Funds") equal to one-twelfth of: (a) yearly taxes and assessments which may attain priority over this Security Instrument; (b) yearly leasehold payments or ground rents on the Property, if any; (c) yearly hazard insurance premiums; and (d) yearly mortgage insurance premiums, if any. These items are called "escrow items." Lender may estimate the Funds due on the basis of current data and reasonable estimates of future escrow items.

The Funds shall be held in an institution the deposits or accounts of which are insured or guaranteed by a federal or state agency (including Lender if Lender is such an institution). Lender shall apply the Funds to pay the escrow items. Lender may not charge for holding and applying the Funds, analyzing the account or verifying the escrow items, unless Lender pays Borrower interest on the Funds and applicable law permits Lender to make such a charge. Borrower and Lender may agree in writing that interest shall be paid on the Funds. Unless an agreement is made or applicable law requires interest to be paid, Lender shall not be required to pay Borrower any interest or earnings on the Funds. Lender shall give to Borrower, without charge, an annual accounting of the Funds showing credits and debits to the Funds and the purpose for which each debit to the Funds was made. The Funds are pledged as additional security for the sums secured by this Security Instrument.

If the amount of the Funds held by Lender, together with the future monthly payments of Funds payable prior to the due dates of the escrow items, shall exceed the amount required to pay the escrow items when due, the excess shall be, at Borrower's option, either promptly repaid to Borrower or credited to Borrower on monthly payments of Funds. If the amount of the Funds held by Lender is not sufficient to pay the escrow items when due, Borrower shall pay to Lender any amount necessary to make up the deficiency in one or more payments as required by Lender.

Upon payment in full of all sums secured by this Security Instrument, Lender shall promptly refund to Borrower any Funds held by Lender. If under paragraph 19 the Property is sold or acquired by Lender, Lender shall apply, no later than immediately prior to the sale of the Property or its acquisition by Lender, any Funds held by Lender at the time of application as a credit against the sums secured by this Security Instrument.

3. Application of Payments. Unless applicable law provides otherwise, all payments received by Lender under paragraphs 1 and 2 shall be applied: first, to late charges due under the Note; second, to prepayment charges due under the Note; third, to amounts payable under paragraph 2; fourth, to interest due; and last, to principal due.

4. Charges; Liens. Borrower shall pay all taxes, assessments, charges, fines and impositions attributable to the Property which may attain priority over this Security Instrument, and leasehold payments or ground rents, if any. Borrower shall pay these obligations in the manner provided in paragraph 2, or if not paid in that manner, Borrower shall pay them on time directly to the person owed payment. Borrower shall promptly furnish to Lender all notices of amounts to be paid under this paragraph. If Borrower makes these payments directly, Borrower shall promptly furnish to Lender receipts evidencing the payments.

Borrower shall promptly discharge any lien which has priority over this Security Instrument unless Borrower: (a) agrees in writing to the payment of the obligation secured by the lien in a manner acceptable to Lender; (b) contests in good faith the lien subject to a lien which may attain priority over this Security Instrument, Lender may give Borrower a notice identifying the lien. Borrower shall satisfy the lien or take one or more of the actions set forth above within 10 days of the giving of notice.

5. Hazard Insurance. Borrower shall keep the improvements now existing or hereafter erected on the Property insured against loss by fire, hazards included within the term "extended coverage," and any other hazards for which Lender requires insurance. This insurance shall be maintained in the amounts and for the periods that Lender requires. The insurance carrier providing the insurance shall be chosen by Borrower subject to Lenders' approval which shall not be unreasonably withheld.

All insurance policies and renewals shall be acceptable to Lender and shall include a standard mortgage clause. Lender shall have the right to hold the policies and renewals. If Lender requires, Borrower shall promptly give to Lender all receipts of paid premiums and renewal notices. In the event of loss, Borrower shall give prompt notice to the insurance carrier and Lender. Lender may make proof of loss if not made promptly by Borrower.

Unless Lender and Borrower otherwise agree in writing, insurance proceeds shall be applied to restoration or repair of the Property damaged, if the restoration or repair is economically feasible and Lender's security is not lessened. If the restoration or repair is not economically feasible or Lender's security would be lessened, the insurance proceeds shall be applied to the sums secured by this Security Instrument, whether or not then due, with any excess paid to Borrower. If Borrower abandons the Property, or does not answer within 30 days a notice from Lender that the insurance carrier has offered to settle a claim, then Lender may collect the insurance proceeds. Lender may use the proceeds to repair or restore the Property or to pay sums secured by this Security Instrument, whether or not then due. The 30-day period will begin when the notice is given.

Unless Lender and Borrower otherwise agree in writing, any application of proceeds to principal shall not extend or postpone the due date of the monthly payments referred to in paragraphs 1 and 2 or change the amount of the payments. If under paragraph 19 the Property is acquired by Lender, Borrower's right to any insurance policies and proceeds resulting from damage to the Property prior to the acquisition shall pass to Lender to the extent of the sums secured by this Security Instrument immediately prior to the acquisition.

6. Preservation and Maintenance of Property; Leaseholds. Borrower shall not destroy, damage or substantially change the Property, allow the Property to deteriorate or commit waste. If this Security Instrument is on a leasehold, Borrower shall comply with the provisions of the lease, and if Borrower acquires fee title to the Property, the leasehold and fee title shall not merge unless Lender agrees to the merger in writing.

7. Protection of Lender's Rights in the Property; Mortgage Insurance. If Borrower fails to perform the covenants and agreements contained in this Security Instrument, or there is a legal proceeding that may significantly affect Lender's rights in the Property (such as a proceeding in bankruptcy, probate, for condemnation or to enforce laws or regulations), then Lender may do and pay for whatever is necessary to protect the value of the Property and Lender's rights in the Property. Lender's actions may include paying any sums secured by a lien which has priority over this Security Instrument, appearing in court, paying reasonable attorneys' fees and entering on the Property to make repairs. Although Lender may take action under this paragraph 7, Lender does not have to do so.

Any amounts disbursed by Lender under this paragraph 7 shall become additional debt of Borrower secured by this Security Instrument. Unless Borrower and Lender agree to other terms of payment, these amounts shall bear interest from the date of disbursement at the Note rate and shall be payable, with interest, upon notice from Lender to Borrower requesting payment.

If Lender required mortgage insurance as a condition of making the loan secured by this Security Instrument, Borrower shall pay the premiums required to maintain the insurance in effect until such time as the requirement for the insurance terminates in accordance with Borrower's and Lender's written agreement or applicable law.

8. Inspection. Lender or its agent may make reasonable entries upon and inspections of the Property. Lender shall give Borrower notice at the time of or prior to an inspection specifying reasonable cause for the inspection.

9. Condemnation. The proceeds of any award or claim for damages, direct or consequential, in connection with any condemnation or other taking of any part of the Property, or for conveyance in lieu of condemnation, are hereby assigned and shall be paid to Lender.

In the event of a total taking of the Property, the proceeds shall be applied to the sums secured by this Security Instrument, whether or not then due, with any excess paid to Borrower. In the event of a partial taking of the Property, unless Borrower and Lender otherwise agree in writing, the sums secured by this Security Instrument shall be reduced by the amount of the proceeds multiplied by the following fraction: (a) the total amount of the sums secured immediately before the taking, divided by (b) the fair market value of the Property immediately before the taking. Any balance shall be paid to Borrower.

If the Property is abandoned by Borrower, or if, after notice by Lender to Borrower that the condemnor offers to make an award or settle a claim for damages, Borrower fails to respond to Lender within 30 days after the date the notice is given, Lender is authorized to collect and apply the proceeds, at its option, either to restoration or repair of the Property or to the sums secured by this Security Instrument, whether or not then due.

Unless Lender and Borrower otherwise agree in writing, any application of proceeds to principal shall not extend or postpone the due date of the monthly payments referred to in paragraphs 1 and 2 or change the amount of such payments.

10. Borrower Not Released; Forbearance By Lender Not a Waiver. Extension of the time for payment or modification of amortization of the sums secured by this Security Instrument granted by Lender to any successor in interest of Borrower shall not operate to release the liability of the original Borrower or Borrower's successors in interest. Lender shall not be required to commence proceedings against any successor in interest or refuse to extend time for payment or otherwise modify amortization of the sums secured by this Security Instrument by reason of any demand made by the original Borrower or Borrower's successors in interest. Any forbearance by Lender in exercising any right or remedy shall not be a waiver of or preclude the exercise of any right or remedy.

11. Successors and Assigns Bound; Joint and Several Liability; Co-signers. The covenants and agreements of this Security Instrument shall bind and benefit the successors and assigns of Lender and Borrower, subject to the provisions of paragraph 17. Borrower's covenants and agreements shall be joint and several. Any Borrower who co-signs this Security Instrument but does not execute the Note: (a) is co-signing this Security Instrument only to mortgage, grant and convey that Borrower's interest in the Property under the terms of this Security Instrument; (b) is not personally obligated to pay the sums secured by this Security Instrument; and (c) agrees that Lender and any other Borrower may agree to extend, modify, forbear or make any accommodations with regard to the terms of this Security Instrument or the Note without that Borrower's consent.

12. Loan Charges. If the loan secured by this Security Instrument is subject to a law which sets maximum loan charges, and that law is finally interpreted so that the interest or other loan charges collected or to be collected in connection with the loan exceed the permitted limits, then: (a) any such loan charge shall be reduced by the amount necessary to reduce a direct payment to Borrower. If a refund reduces principal, the reduction will be treated as a partial prepayment without any prepayment charge under the Note.

13. Legislation Affecting Lender's Rights. If enactment or expiration of applicable laws has the effect of rendering any provision of the Note or this Security Instrument unenforceable according to its terms, Lender, at its option, may require immediate payment in full of all sums secured by this Security Instrument and may invoke any remedies permitted by paragraph 19. If Lender exercises this option, Lender shall take the steps specified in the second paragraph of paragraph 17.

14. Notices. Any notice to Borrower provided for in this Security Instrument shall be given by delivering it or by mailing it by first class mail unless applicable law requires use of another method. The notice shall be directed to the Property Address or any other address Borrower designates by notice to Lender. Any notice to Lender shall be given by first class mail to Lender's address stated herein or any other address Lender designates by notice to Borrower. Any notice provided for in this Security Instrument shall be deemed to have been given to Borrower or Lender when given as provided in this paragraph.

15. Governing Law; Severability. This Security Instrument shall be governed by federal law and the law of the jurisdiction in which the Property is located. In the event that any provision or clause of this Security Instrument or the Note conflicts with applicable law, such conflict shall not affect other provisions of this Security Instrument or the Note which can be given effect without the conflicting provision. To this end the provisions of this Security Instrument and the Note are declared to be severable.

16. Borrower's Copy. Borrower shall be given one conformed copy of the Note and of this Security Instrument.

17. Transfer of the Property or a Beneficial Interest in Borrower. If all or any part of the Property or any interest in it is sold or transferred (or if a beneficial interest in Borrower is sold or transferred and Borrower is not a natural person) without Lender's prior written consent, Lender may, at its option, require immediate payment in full of all sums secured by this Security Instrument. However, this option shall not be exercised by Lender if exercise is prohibited by federal law as of the date of this Security Instrument.

If Lender exercises this option, Lender shall give Borrower notice of acceleration. The notice shall provide a period of not less than 30 days from the date the notice is delivered or mailed within which Borrower must pay all sums secured by this Security Instrument. If Borrower fails to pay these sums prior to the expiration of this period, Lender may invoke any remedies permitted by this Security Instrument without further notice or demand on Borrower.

18. Borrower's Right to Reinstate. If Borrower meets certain conditions, Borrower shall have the right to have enforcement of this Security Instrument discontinued at any time prior to the earlier of : (a) 5 days (or such other period as applicable law may specify for reinstatement) before sale of the Property pursuant to any power of sale contained in this Security Instrument; or (b) entry of a judgment enforcing this Security Instrument. Those conditions are that the Borrower: (a) pays Lender all sums which then would be due under this Security Instrument and the Note had no acceleration occurred; (b) cures any default of any other covenants or agreements; (c) pays all expenses incurred in enforcing this Security Instrument, including, but not limited to, reasonable attorneys' fees; and (d) takes such action as Lender may reasonably require to assure that the lien of this Security Instrument, Lender's rights in the Property and Borrower's obligation to pay the sums secured by this Security Instrument shall continue unchanged. Upon reinstatement by Borrower, this Security Instrument and the obligations secured hereby shall remain fully effective as if no acceleration had occurred. However, this right to reinstate shall not apply in the case of acceleration under paragraphs 13 or 17.

NON-UNIFORM COVENANTS. Borrower and Lender further covenant and agree as follows:

19. **Acceleration; Remedies.** Lender shall give notice to Borrower prior to acceleration following Borrower's breach of any covenant or agreement in this Security Instrument (but not prior to acceleration under paragraphs 13 and 17 unless applicable law provides otherwise). The notice shall specify: (a) the default; (b) the action required to cure the default; (c) a date, not less than 30 days from the date the notice is given to Borrower, by which the default must be cured; and (d) that failure to cure the default on or before the date specified in the notice may result in acceleration of the sums secured by this Security Instrument and sale of the Property. The notice shall further inform Borrower of the right to reinstate after acceleration and the right to bring a court action to assert the non-existence of a default or any other defense of Borrower to acceleration and sale. If the default is not cured on or before the date specified in the notice, Lender at its option may require immediate payment in full of all sums secured by this Security Instrument without further demand and may invoke the power of sale and any other remedies permitted by applicable law. Lender shall be entitled to collect all expenses incurred in pursuing the remedies provided in this paragraph 19, including, but not limited to, reasonable attorneys' fees and costs of title evidence.

If Lender invokes the power of sale, Lender or Trustee shall give to Borrower (and the owner of the Property, if a different person) notice of sale in the manner prescribed by applicable law. Trustee shall give public notice of sale by advertising, in accordance with applicable law, once a week for two successive weeks in a newspaper having general circulation in the county or city in which any part of the Property is located, and by such additional or any different form of advertisement the Trustee deems advisable. Trustee may sell the Property on the eighth day after the first advertisement or any day thereafter, but not later than 30 days following the last advertisement. Trustee, without demand on Borrower, shall sell the Property at public auction to the highest bidder at the time and place and under the terms designated in the notice of sale in one or more parcels and in any order Trustee determines. Trustee may postpone sale of all or any parcel of the Property by advertising in accordance with applicable law. Lender or its designee may purchase the Property at any sale.

Trustee shall deliver to the purchaser Trustee's deed conveying the Property with special warranty of title. The recitals in the Trustee's deed shall be prima facie evidence of the truth of the statements made therein. Trustee shall apply the proceeds of the sale in the following order: (a) to all expenses of the sale, including, but not limited to, Trustee's fees of ......% of the gross sale price and reasonable attorneys' fees; (b) to the discharge of all taxes, levies and assessments on the Property, if any, as provided by applicable law; (c) to all sums secured by this Security Instrument; and (d) any excess to the person or persons legally entitled to it. Trustee shall not be required to take possession of the Property prior to the sale thereof or to deliver possession of the Property to the purchaser at the sale.

20. **Lender in Possession.** Upon acceleration under paragraph 19 or abandonment of the Property, Lender (in person, by agent or by judicially appointed receiver) shall be entitled to enter upon, take possession of and manage the Property and to collect the rents of the Property including those past due. Any rents collected by Lender or the receiver shall be applied first to

release this Security Instrument without charge to Borrower. Borrower shall pay any recordation costs.

22. Substitute Trustee. Lender, at its option, may from time to time remove Trustee and appoint a successor trustee to any trustee appointed hereunder. Without conveyance of the Property, the successor trustee shall succeed to all the title, power and duties conferred upon Trustee herein and by applicable law.

23. Identification of Note. The Note is identified by a certificate on the Note executed by any Notary Public who certifies an acknowledgment hereto.

24. Riders to this Security Instrument. If one or more riders are executed by Borrower and recorded together with this Security Instrument, the covenants and agreements of each such rider shall be incorporated into and shall amend and supplement the covenants and agreements of this Security Instrument as if the rider(s) were a part of this Security Instrument. [Check applicable box(es)]

☐ Adjustable Rate Rider   ☐ Condominium Rider   ☐ 2–4 Family Rider

☐ Graduated Payment Rider   ☐ Planned Unit Development Rider

☐ Other(s) [specify]

NOTICE: THE DEBT SECURED HEREBY IS SUBJECT TO CALL IN FULL OR THE TERMS THEREOF BEING MODIFIED IN THE EVENT OF SALE OR CONVEYANCE OF THE PROPERTY CONVEYED.

BY SIGNING BELOW, Borrower accepts and agrees to the terms and covenants contained in this Security Instrument and in any rider(s) executed by Borrower and recorded with it.

....(Seal)
—Borrower

....(Seal)
—Borrower

——————————— [Space Below This Line For Acknowlegment] ———————————

Let's say the lender is owed $50,000 at the time of the foreclosure and the property brings $65,000. What happens to the difference of $15,000? The difference goes to the owner. But if the sale brings in only $35,000, who pays the difference? The owner, if he can. If he cannot, he still owes $15,000 plus attorney's fees.

Of course, most properties will never go to foreclosure even if there is trouble meeting payments. When someone is slow making payments, usually the lender begins calling and asking, "Hey, Charlie, will you start paying?"

"Yeah, I'll start paying," says Charlie. And then Charlie is slow again.

"Will you start paying?"

"Yeah."

At this point the loan is costing the lender a great deal of money and administrative effort because the law will not permit him to have a delinquent note unless he is doing something about it. So finally the lender says, "Charlie, you're never going to pay this. I'm going to accelerate and go to foreclosure."

When Charlie sees his name in the paper he realizes that the lender is not playing games. The choice facing Charlie is either to pay the note in full, or lose the property to foreclosure. If he cannot make the payments he must sell his house.

Keep your payments current. If you're having difficulty making them, discuss your situation honestly and realistically with the lender. Do not ignore the problem. It won't go away.

F. PREPAYMENT PENALTY

A prepayment penalty is one that is assessed when you pay off your loan before its maturity date. Assume you bought a house and took out a twenty-year mortgage, then sold it after three or four years (as many people do). A few years ago lenders sometimes imposed a penalty if you prepaid the loan, and sometimes these penalties were quite stiff. Today most states have outlawed prepayment penalties, or else severely limited the time period during which there can be a penalty. You should check to see if there is a prepayment penalty in your note. If there is, evaluate the penalty in making your decision to accept the loan. However, you need not reject an otherwise good loan because of this provision alone.

G. RIGHT OF ANTICIPATION RESERVED

The right of anticipation is a legal term. It means that you have the right to prepay without penalty. So the term might appear to be against you, but it really is for your benefit.

If your loan doesn't have a prepayment penalty, you will frequently see the legal term "right of anticipation reserved" on either the note, the deed of trust (mortgage), or both. Literally, it means the right to prepay (anticipate) is available (reserved) to the note maker (borrower).

H. ASSUMPTION

The right of assumption means that if you sell your house you can permit the person who buys it to assume your loan. Most people assume they have this right, and if there is nothing in the deed of trust to prevent this, then it is a right. However, some deeds of trust will specifically deny you this right. Sometimes it is specified that assumption will be allowed providing that the person assuming the loan can comply with the normal credit risk requirements. Having a loan which is assumable is a preferred position, but not critical.

I. ACKNOWLEDGEMENT AND RECORDATION

The reason that instruments are acknowledged is so that they can be recorded. The courts want them recorded because it puts the world on notice of whatever the instrument purports to tell—a deed, a deed of trust, a lien, or other document. This is called "constructive notice." Anything you record in a clerk's office has to be notarized.

There are several reasons why recording is important. Recordation puts the world on notice that a transaction involving this property has taken place.

Suppose you purchased a piece of property called Timberwood from Sam for $50,000 in cash. You give Sam the $50,000 and Sam gives you a deed to Timberwood. (Your first problem may be that Sam either does not own the property or that if he does he still owes money on it.) But let's assume he owns the property free and clear. If you don't record this deed, then the world, including Charlie, does not know you own Timberwood. A few weeks later Sam decides to sell the property to Charlie for $75,000. You know that Sam does not own Timberwood, but Charlie and Charlie's attorney do not. In fact, when Charlie's attorney checks the title at the Court House he will find that Sam, not you, is still on record as owning Timberwood. Charlie now gives Sam $75,000, takes another deed from Sam, and records it. So who owns Timberwood? Charlie. Because Charlie bought it in good faith without notice that you purchased it first. Sam was paid twice, and you can sue Sam for what he did, but Charlie owns the property, and there's nothing you can do about it.

Suppose you recorded your deed to Timberwood immediately after Sam gave it to you. Then suppose Charlie approached you a week later and offered you $75,000 for Timberwood. You agree to sell—you will make $25,000 in a week—but you cannot find the original deed. Do you still own Timberwood? Yes, because you recorded the deed, and now the whole world knows that you are the proud owner of Timberwood. The original deed is nice to have, but once it is recorded it is not especially important. The moral is: Record your instrument as soon as possible. A recorded document is notice to the world. It's your ultimate protection.

J. RELEASES

When you pay off your note, make sure that you obtain the paid note. Have your attorney take it down to the clerk's office and have your deed of trust released by an instrument called a "certificate of satisfaction." People frequently forget to do this. They'll say, "I paid off my note. Everything is fine." In all probability it is, but in the bureaucracy of America the paid-up note could get lost. If you do not obtain your certificate of satisfaction, it doesn't mean that you will be held liable for the note. But you might develop a real headache when you go to sell or refinance your property.

The documents pertaining to your property are important, and we hope this chapter has given you a better understanding of them. But don't try to be your own lawyer. When it comes to dealing with these documents and the legal process trust your lawyer and your instincts. And don't forget to purchase owner's title insurance.

12 Warranties

When you build your home you'll be confronted with a variety of warranties, ranging from a thirty-year guarantee on treated wood to a five-year warranty on your air-conditioning compressor, or a ninety-day warranty on some electrical appliances. Warranties come in all types, sizes, and durations and sometimes can be confusing. It's important to study all your warranties carefully, especially those covering major items. Some warranties don't give you as complete a guarantee as you may think. On some equipment the duration of the warranties on different parts will vary. For example, an air-conditioning compressor might be guaranteed for five years, while other parts in the heating/air conditioning unit are guaranteed for only a year.

A. TYPES

1. Express

Express warranties are usually written out, although they don't have to be. They spell out the conditions and limits of the warranty in detail, so that there can be no misunderstanding about what the manufacturer guarantees.

2. Implied

An implied warranty is created by actions rather than words, and in almost every case it exists because the state or federal government has imposed a warranty upon the vendor, the mechanic, the material man, the manufacturer, the supplier, and so on.

You buy a toaster manufactured by El Cheapo Toaster Company. You take the toaster home and it doesn't toast. Do you have a warranty? Yes. You have one even if the El Cheapo Toaster Company is in Singapore and no warranty is printed on the box. There is a warranty because the law, under the Uniform Commercial Code, states that a product must suit the need for which it was sold. If you buy a toaster and it doesn't toast, then it doesn't suit the need for which it was sold. Most states operate under the Uniform Commercial Code or some other commercial code or statute that provides the consumer with some type of protection. This warranty is called specifically the warranty of merchantability and it applies primarily to manufactured goods rather than to housing.

Housing is not specifically covered by the Uniform Commercial Code. But it is covered, we believe, in every state by another statutory provision which specifies that housing must meet the use for which it was sold. Under this statutory provision for warranty, a new house must be structurally sound and reasonably and substantially meet the use for which it was sold. This does not mean that if you dislike the wallpaper you can cite the state statute and change the wallpaper. What this statute says is that the house must be built in accordance with the codes, and can be used as a house.

In most states, the statutory warranty for a new home is one year, although this may vary from state to state. Be sure to check the statutes of your state. If you buy an existing home, you don't have an implied warranty.

Sometimes you'll see in a contract—either for the sale of a toaster or the sale of a home—that a product is sold without any warranties. But in the case of the home (and usually the toaster) you still have a warranty, because the builder cannot, by contract, make you waive his obligation under the statute. You cannot contract away your rights granted to you by law.

Here is an example. You go into a dry cleaner's, and you see a sign that says: "Our responsibility for damage to your clothing will not exceed ten times the cleaning price." It costs $1.25 to clean the shirt, but you may have spent $30, $40, even $60 to buy the shirt. The cleaner negligently either loses or ruins your $60 shirt. What is the extent of the dry cleaner's liability? The cleaner is liable for the full value of the shirt because he agreed to take your shirt and clean it. If he's doing it for money he has an obligation to return the shirt to you in essentially the same condition that you gave it to him. Neither he nor you can contract away that right.

Why does the dry cleaner put that sign up? Because it is smart business. Most people will just take the offered amount and run. A home builder, of course, will not get away so easily, because new homes cost a great deal more than new shirts.

3. Warranties for Homeowners

In addition to the implied and express warranties that will come with your house, there are two types of warranties available that apply specifically to your house.

a. THIRD-PARTY HOMEOWNER'S WARRANTY

In a third-party warranty, you are the first party, the builder is the second party, and the third party is usually an insurance company. This kind of warranty is essentially an insurance policy which you can buy in the same way that you buy title insurance or fire insurance. It warrants certain components of your home for certain periods of time.

Third-party warranties are tricky. They usually run for ten years, although this ten-year protection is what lawyers call a "fallacy of composition," because all of the elements of the home are not covered for ten years in the warranty. What the warranty means is that you have up to ten years' coverage on parts of the home. When you read a third-party warranty carefully, it will describe specifically what is covered, and for what length of time. Usually the structural components of your house are covered for ten years; windows may be covered for five years; appliances may be covered for two or three, leaky basements for a year, and so on down the line. Third-party warranties are usually written by regular insurance underwriters.

b. TWO-PARTY WARRANTY

There are also large builders who have their own warranty program. A builder may decide that he can offer a warranty program which is as cost effective to both the customer and to him as one which is offered by a third party. In some circumstances you may want to accept the warranty from a builder. It may be for a longer term, better, or cheaper.

There are also some counties that are now requiring some warranties beyond the statuary one-year warranties. Builders are being required to offer some type of warranty, or third-party warranties, on their homes so that their customers will protected.

b. WARRANTY LIMITS

1. Full and complete

Occasionally, you'll find an unconditional, no-hassle warranty. We sometimes shop at a building supply store that generally has higher prices than its competitors, but offers an unconditional warranty for most of its merchandise. "If you buy an item from us and you don't like it," they say, "bring it back and we will repair it or replace it for free." That's a no-limitation, hassle-free type of warranty. Rarely if ever will you see a warranty like that in the housing industry. Most building warranties are limited. But a limited warranty is by no means worthless.

2. Limited

The federal government has imposed a statute which says that any warranty that is not full and unconditional must state that it is a limited warranty. Most of the warranties you will be given, whether for your toaster or your house, will be limited warranties. But don't be concerned. This simply means that it is not unconditional. If your warranty is limited, and this fact is not stated in bold print as required be federal statute, you can argue the point.

3. What Does the Warranty Cover?

The wording of a warranty can be deceiving. When they give you an unconditional, lifetime warranty on your automobile muffler, it means that your muffler itself is warranted. It does not mean that the cost of labor necessary to replace the muffler is warranted. Nor does it mean that the exhaust system is warranted, or the clamps or the bolts.

4. Consequential Damages

Suppose you buy a muffler and it is installed improperly and your car blows up, which also damages the undercarriage. Is the company that you bought the muffler from responsible for the damage to your car as a result of this negligent installation or negligent manufacture of the muffler? Absolutely and categorically yes. Suppose, however, that the muffler wears out over a period of time, and as it does, it also wears out the exhaust system on the car. Is the muffler company responsible for replacing the exhaust system? No, because the muffler degenerated in normal use over time.

How does this relate to the building of your home? When your plumber warrants the pipes for a period of five years, and the pipe wears out not because of negligent installation or construction but because of the water running through it over a normal period of time, and it leaks and soaks up your drywall and your insulation, is the plumber responsible for the insulation and the drywall? Absolutely not. Is he responsible for the repair of the pipe itself if the warranty time has not run out? Yes.

Most warranties run only to the direct liability—the direct damage associated with the malfunctioning or wearing out of the particular item. They do not also cover consequential damages. One company we know gives a warranty for the structural components of a house which runs for fifteen years. If there is a problem with the structural component itself, the company will repair it. But if the drywall has to be removed in order to repair the faulty structure, the company will not pay for the removal of the drywall, because according to the warranty it must have free and easy access to the problem covered by the warranty.

5. Conditions

Under contract law, there are conditions precedent and conditions subsequent. Here are examples and an explanation of what these legal terms mean:

a. PRECONDITION (CONDITION PRECEDENT)

Frequently warranties are given on the basis of: "Buy during the month of June and get an extended warranty." In order to get that extended warranty you must, in fact, buy during the month of June. That is a precondition necessary to obtaining the warranty.

"Buy with your American Express card and get an extended warranty." That's another precondition. A precondition for a warranty that's very prevalent in todays' market is: "Pay more money and get a warranty." You can in fact pay money for a warranty, and that's a precondition to some warranties.

b. SUBSEQUENT CONDITION

A warranty is only valid if you use the product for the purpose and use for which it was intended. If you buy a Yugo, enter it in the Indianapolis 500, and go through two hundred laps, you can't bring it back and say it didn't work. You can't put six tons of concrete block on a one-ton pickup truck and take the truck back and say it didn't work. A classic example of subsequent condition is that you must keep the oil in your car fully replenished and clean for your engine warranty to be valid.

Subsequent conditions can take many different forms. You buy a lawn mower and receive a warranty with the lawn mower. The warranty requires you to oil the lawn mower and you do not do so. Is the warranty voided? Absolutely. These are conditions subsequent to a warranty that you must comply with in order to maintain that warranty. Most warranties will say that they only apply if you use whatever it is they cover—your house, your lawn mower—under normal conditions and with proper care. Some warranties, however, will go beyond that and will give you a warranty even though you do something which is above and beyond normal use. For example, some car dealers will offer a warranty stating that if you use a specific oil for the life of your car, and you oil your car with that product in accordance with a schedule specified by the dealer, they will warrant the engine of your car for 100,000 miles. But you must use that specific brand of oil, you must follow the schedule, and you must produce proof that you did both.

That is another kind of subsequent condition warranty that can affect your new home.

6. Transferability

A warranty may or may not be transferable. Check this carefully, whether the item in question is a product or a house.

The third-party warranty is usually transferable from one owner to another. Most warranties associated with a house, however, are not transferable. The reason is strictly economic. Some companies will offer warranties for as long as fifteen years, but this isn't as dramatic as it sounds, because most structural problems with a house occur within two years after construction. Thus warranting something from the third year to the fifteenth year is a marginal expense. However, the length of time can be relevant, because most houses are sold within eight years after they have been built. So in most cases if a builder offers a fifteen year warranty he is more than likely committing himself to only eight years, if he stipulates that the warranty is non-transferable, because the house will probably be sold in about eight years.

C. STABILITY OF THE ISSUER

Finally, one of the most important considerations in assessing a warranty is the stability and reliability of the warrantor. If a major builder offers you a comfortable warranty then you probably have a good deal because more than likely he'll be in a position to back it up. Sometimes a builder will also pass along warranties to you from various manufacturers—the Andersen window warranty, the General Electric warranty on your appliances, or other. These warranties should be kept for subsequent use.

Warranties should not be taken lightly; you should study them carefully. Insist on warranties where you feel they are traditional or justified, and be sure to keep good records. You probably won't need them until one to five years later, and they're easy to lose or misplace. Before a company will honor a warranty, it will more than likely insist that you prove the warranty is yours, that it was properly earned, you met the preconditions, and you met the subsequent conditions.

13 CONCLUSION

We've given you the basic information you need to build your own home. Obviously it involves a lot of time and attention. It's hard work and often frustrating. But it can also be fun, rewarding, challenging, and profitable. Do the benefits outweigh the costs? We feel that the answer for most people is yes. Expect the unexpected, don't underestimate cost and time frames, and don't forget that tasks done quickly and properly reap big rewards.

A. THE MOST COMMON MISTAKES

The first-time homebuilder will inevitably make some mistakes. Here are the six biggest mistakes homebuilders make, which you should try to avoid:

1. Accepting the Low Bid

It is always tempting to accept the low bid. It may be the right thing to do providing you take certain precautions. However, be aware of the following:

(1) Sometimes the "low bidder" is surprised to hear he is the low bidder. Why? Because he guessed at his bid or just didn't know how to bid.

(2) The low bid is only valuable if the bidder can perform. Most non-performance comes from the low bidder.

(3) The low bid may be the result of the bidder either intentionally or unintentionally leaving something out of the bid.

(4) Some low bidders plan on "hitting you up" for additional charges as you go along.

(5) If a low bidder cannot perform, there is very little you can do from a practical standpoint. The legal cost would exceed the benefit.

In summary, make sure that the bidder is knowledgeable and trustworthy. If his bid is the lowest, you're okay. If not, select knowledge and honesty over price.

2. Choosing Price Over Quality

In most cases if your home plan does not match your budget, our recommendation is to downsize the project rather than the specifications. A quality, well-built home is a joy forever. An inexpensive home is neither a thing of beauty nor a wise investment.

3. Downsizing Rooms To Save Money

In most cases the flow, functionality, and charm of the home is destroyed. It is better to choose a smaller home or delete or defer something like the garage, decks, etc.

4. Analyzing the Cost Value of the Home on the Square-Foot Basis

This is one of the biggest mistakes made. The value of a custom home cannot be measured or compared on a square-foot basis.

5. Operating Without Signed Change Orders

This is the biggest single source of misunderstanding. Keep your arrangements with your contractors safe and simple. Write them down, specifying a price, with both of you understanding and signing the change order.

6. Acting Against Your Comfort Level

Perhaps the biggest mistake you can make in the entire building process is acting against your comfort level. If you are not comfortable with the contractor, supplier, lender, or anyone else, his specifications, his price, or his promise may mean nothing. Make sure you have a comfortable feeling about the person you are dealing with and what he is offering to do. In the long run, his legitimacy and your working relationship is more important than the price. A low price is of no value if there is unacceptable performance.

Again, although we do not recommend you do any of the construction work yourself, we want to emphasize that anyone who has the creative urge to undertake some aspect of the actual building should not hesitate to do so, even if it might cost you a little more. Doing any part of the actual work, from installing built-in bookcases to trimming and painting a room, is all part of the fun and satisfaction you gain from building your own home. Included in the reading list at the end of this book are several books designed to assist you on most of the actual hands-on construction work that needs to be done on a house.

One of the Greek philosophers said that every man should build his own home at least once. There is probably no more satisfying experience in life than doing something that is not only fun but financially rewarding. And most people we know who have built their home not only say it was one of the most memorable periods of their life but are ready to do it again. One reason was that the house they built themselves has already doubled or tripled in value. In fact, professional homebuilders spend their lives building homes and making money. And we don't believe we have ever met a homebuilder who did not enjoy what he was doing. In short, it is a rewarding experience even building somebody else's home. Building your own home—your castle—is especially rewarding.

SUGGESTED READING

Alth, Max & Charlotte. *Be Your Own Contractor: The Affordable Way To Home Ownership.* Tab Books, Inc.: Blue Ridge Summit, Pennsylvania, 1984

Coffee, Frank. *Everything You Need To Know About Creative Home Financing, New Affordable Ways to Buy (And Sell) A Home, Condo, or Co-Op.* Simon and Schuster: New York, 1982

Conran, Terence. *The House Book.* Crown Publishers: New York, 1982

DiDonno, Lupe; Sperling, Phyllis. *How to Design and Build Your Own House.* Alfred A. Knopf: New York, 1986

Galaty, Fillmore W.; Allaway, Wellington J.; & Kyle, Robert C. *Modern Real Estate Practice, 10th Edition.* Real Estate Education Company: Chicago, 1985

Gilmore, Louis. *For Sale By Owner: How To Sell Your Own Home Without A Broker and Save Thousands of Dollars.* Simon and Schuster: New York, 1980.

Greenfield, Ellen J. *House Dangerous.* Random House: New York, 1987.

Hall, Barbara Jane. *101 Easy Ways To Make Your Home Sell Faster.* Ballentine Books: New York, 1985

Harrison, Henry S. *Houses (The Illustrated Guide to Construction, Design, and Systems).* National Association of Realtors: 1973

Hasenau, J. James. *Build Your Own Home.* Holland House Press: Michigan, 1977

Heldman, Carl. *How To Save 25% Without Lifting A Hammer—Be Your Own House Contractor.* Garden Way Publishing: Pownal, Vermont, 1986

Jackson, W. P. *Estimating Home Building Costs.* Craftsman Book Company: Carlsbad, California, 1981

Janik, Carolyn & Rejnis, Ruth. *All America's Real Estate Book: Everyone's Guide to Buying, Selling, Renting and Investing.* Penguin Books: New York, 1985

Karrass, Dr. Chester. *Effective Negotiating: Workbook and Discussion Guide.* Karras Seminars: California

Nisson, J. D. Ned & Dutt, Gautam. *The Superinsulated Home Book.* John Wiley & Sons: New York, 1985.

Orman, Halsey Van. *Illustrated Handbook of Home Construction.* Van Nostrand Reinholt Company: New York, 1982

Roskind, Robert. *Before You Build (A Preconstruction Guide).* Ten Speed Press: Berkeley, California, 1985

Scutella, R. M. & Heberle, David. *How To Plan, Contract and Build Your Own Home.* Tab Books, Inc.: Blue Ridge Summit, Pennsylvania, 1987

Syvanen, Bob. *What It's Like to Build A House: The Diary of A Builder.* The Taunton Press: Newtown, Connecticut, 1985

Wagner, Willis H. *Modern Carpentry (Building Construction Details in Easy-To-Understand Form).* Goodheart-Willcox Company: South Holland, Illinois, 1983

Watkins, A. M. *How to Avoid The 10 Biggest Home-Buying Traps.* The Building Institute, 1984

Watkins, A. M. *How to Avoid The 10 Biggest Home-Buying Traps.* The Building Institute, 1985

Watson, Donald. *Designing and Building a Solar House.* Garden Way Publishing: Pownal, Vermont, 1985

Index

Numbers in bold face are documents